Nanocolloids for Petroleum Engineering

Nanocolloids for Petroleum Engineering

Fundamentals and Practices

Baghir A. Suleimanov
Oil and Gas Scientific Research Project Institute
State Oil Company of Azerbaijan Republic (SOCAR)
Baku, Azerbaijan

Elchin F. Veliyev
Oil and Gas Scientific Research Project Institute
State Oil Company of Azerbaijan Republic (SOCAR)
Baku, Azerbaijan

Vladimir Vishnyakov
University of Huddersfield
Huddersfield, UK

This edition first published 2022
© 2022 by John Wiley & Sons Ltd. All rights reserved.

All rights reserved. No part of this publication may be reproduced, stored in a retrieval system, or transmitted, in any form or by any means, electronic, mechanical, photocopying, recording or otherwise, except as permitted by law. Advice on how to obtain permission to reuse material from this title is available at http://www.wiley.com/go/permissions.

The right of Baghir A. Suleimanov, Elchin F. Veliyev and Vladimir Vishnyakov to be identified as the authors of this work has been asserted in accordance with law.

Registered Offices
John Wiley & Sons, Inc., 111 River Street, Hoboken, NJ 07030, USA
John Wiley & Sons Ltd, The Atrium, Southern Gate, Chichester, West Sussex, PO19 8SQ, UK

Editorial Office
The Atrium, Southern Gate, Chichester, West Sussex, PO19 8SQ, UK

For details of our global editorial offices, customer services, and more information about Wiley products visit us at www.wiley.com. Wiley also publishes its books in a variety of electronic formats and by print-on-demand. Some content that appears in standard print versions of this book may not be available in other formats.

Limit of Liability/Disclaimer of Warranty
In view of ongoing research, equipment modifications, changes in governmental regulations, and the constant flow of information relating to the use of experimental reagents, equipment, and devices, the reader is urged to review and evaluate the information provided in the package insert or instructions for each chemical, piece of equipment, reagent, or device for, among other things, any changes in the instructions or indication of usage and for added warnings and precautions. While the publisher and authors have used their best efforts in preparing this work, they make no representations or warranties with respect to the accuracy or completeness of the contents of this work and specifically disclaim all warranties, including without limitation any implied warranties of merchantability or fitness for a particular purpose. No warranty may be created or extended by sales representatives, written sales materials or promotional statements for this work. The fact that an organization, website, or product is referred to in this work as a citation and/or potential source of further information does not mean that the publisher and authors endorse the information or services the organization, website, or product may provide or recommendations it may make. This work is sold with the understanding that the publisher is not engaged in rendering professional services. The advice and strategies contained herein may not be suitable for your situation. You should consult with a specialist where appropriate. Further, readers should be aware that websites listed in this work may have changed or disappeared between when this work was written and when it is read. Neither the publisher nor authors shall be liable for any loss of profit or any other commercial damages, including but not limited to special, incidental, consequential, or other damages.

For general information on our other products and services or for technical support, please contact our Customer Care Department within the United States at (800) 762-2974, outside the United States at (317) 572-3993 or fax (317) 572-4002.

Wiley also publishes its books in a variety of electronic formats. Some content that appears in print may not be available in electronic formats. For more information about Wiley products, visit our web site at ww.wiley.com.

Library of Congress Cataloging-in-Publication Data applied for:

Hardback ISBN: 9781119889595

Cover Design: Wiley
Cover Image: © sharply_done

Set in 9.5/12.5pt STIXTwoText by Straive, Pondicherry, India
Printed and bound by CPI Group (UK) Ltd, Croydon, CR0 4YY

Contents

Acknowledgments *ix*
Introduction *xi*

Part A Nanocolloids – An Overview *1*

1	**Nanocolloid Classification** *3*	
1.1	What is a Colloid? *3*	
1.1.1	Colloid Classification *3*	
1.1.2	Colloid Evaluation *4*	
1.2	What is a Nanocolloid? *5*	
2	**Nanocolloid Properties** *7*	
2.1	Different Kinds of Interactions in Nanocolloids *7*	
2.1.1	Van der Waals Interactions *7*	
2.1.2	Electrostatic Interaction *7*	
2.1.3	Elastic–Steric Interaction *8*	
2.1.4	Hydrophobic Interaction *8*	
2.1.5	Solvation Interaction *8*	
2.1.6	Depletion Interaction *8*	
2.1.7	Magnetic Dipole–Dipole Interaction *8*	
2.1.8	Osmotic Repulsion *9*	
2.2	The Stability of Nanocolloids *9*	
2.3	Rheology of Nanocolloids *10*	
2.3.1	Effect of Nanoparticle Interaction on the Colloid Rheology *10*	
2.3.2	Effect of Nanoparticle Migration on the Colloid Rheology *14*	
2.4	Surface Tension. Wettability *21*	
2.4.1	Wettability Alteration *21*	
2.4.2	Surface Tension *23*	

Nomenclature *25*
References *27*

Part B Reservoir Development *33*

3   Reservoir Conditions for Nanocolloid Formation *35*
3.1   In-Situ Formation of Nanogas Emulsions *35*
3.1.1   Stability of the Subcritical Gas Nuclei *35*
3.2   In-Situ Formation of Nanoaerosoles *38*
3.2.1   Stability of the Subcritical Liquid Nuclei *39*

4   Nanogas Emulsions in Oil Field Development *41*
4.1   Hydrodynamics of Nanogas Emulsions *41*
4.1.1   Flow Mechanism of Gasified Newtonian Liquids *41*
4.1.2   Flow of Gasified Newtonian Liquids in Porous Media at Reservoir Conditions *52*
4.2   Hydrodynamics of Nanogas Emulsions in Heavy Oil Reservoirs *60*
4.2.1   Flow Mechanism of Gasified Non-Newtonian Liquids *60*
4.2.2   Flow of Gasified Non-Newtonian Liquids in Porous Media at Reservoir Conditions *66*
4.3   Field Validation of Slippage Phenomena *73*
4.3.1   Steady-State Radial Flow *73*
4.3.2   Unsteady State Flow *86*
4.3.3   Viscosity Anomaly Near to the Phase Transition Point *95*

5   Nanoaerosoles in Gas Condensate Field Development *107*
5.1   Study of the Gas Condensate Flow in a Porous Medium *107*
5.2   Mechanism of the Gas Condensate Mixture Flow *111*
5.2.1   Rheology Mechanism of the Gas Condensate Mixture During Steady-State Flow *112*
5.2.2   Mechanism of Porous Medium Wettability Influence on the Steady-State Gas Condensate Flow *120*
5.2.3   Mechanism of Pressure Build-Up at the Unsteady-State Flow of the Gas Condensate *121*
5.2.4   Concluding Remarks *124*

Nomenclature *125*
References *127*

Part C Production Operations *131*

6   An Overview of Nanocolloid Applications in Production Operations *133*

7   Nanosol for Well Completion *137*
7.1   The Influence of the Specific Surface Area and Distribution of Particles on the Cement Stone Strength *139*
7.2   The Influence of Nano-SiO_2 and Nano-TiO_2 on the Cement Stone Strength *140*
7.3   Regression Equation *141*
7.4   Concluding Remarks *142*

8	**Nanogas Emulsion for Sand Control** *145*
8.1	Fluidization by Gasified Fluids *145*
8.1.1	Carbon Dioxide Gasified Water as Fluidizing Agent *146*
8.1.2	Natural Gas or Air Gasified Water as Fluidizing Agent *149*
8.2	Chemical Additives Impact on the Fluidization Process *151*
8.2.1	Water–Air Mixtures with Surfactant Additives as Fluidizing Agent *151*
8.2.2	Fluidization by Polymer Compositions *152*
8.3	Mechanism of Observed Phenomena *153*

9	**Vibrowave Stimulation Impact on Nanogas Emulsion Flow** *157*
9.1	Exact Solution *158*
9.2	Approximate Solution *161*
9.3	Concluding Remarks *162*

Nomenclature *165*
References *167*

Part D Enhanced Oil Recovery *171*

10	**An Overview of Nanocolloid Applications for EOR** *173*
10.1	Core Flooding Experiments Focused on Dispersion Phase Properties *174*
10.2	Core Flooding Experiments Focused on Dispersion Medium Properties *175*

11	**Surfactant-Based Nanofluid** *177*
11.1	Nanoparticle Influence on Surface Tension in a Surfactant Solution *177*
11.2	Nanoparticle Influence on the Surfactant Adsorption Process *178*
11.3	Nanoparticle Influence on Oil Wettability *179*
11.4	Nanoparticle Influence on Optical Spectroscopy Results *179*
11.5	Nanoparticle Influence on the Rheological Properties of Nanosuspension *182*
11.6	Nanoparticle Influence on the Processes of Newtonian Oil Displacement in Homogeneous and Heterogeneous Porous Mediums *183*
11.7	Concluding Remarks *186*

12	**Nanofluids for Deep Fluid Diversion** *187*
12.1	Pre-formed Particle Nanogels *187*
12.1.1	Nanogel Strength Evaluation *189*
12.1.2	Kinetic Mechanism of Gelation *191*
12.1.3	Core Flooding Experiments *192*
12.1.4	Concluding Remarks *197*
12.2	Colloidal Dispersion Nanogels *197*
12.2.1	Rheology *198*
12.2.2	Aging Effect *200*
12.2.3	Interfacial Tension *200*
12.2.4	Zeta Potential *202*
12.2.5	Particle Size Distribution *203*

| 12.2.6 | Resistance Factor/Residual Resistance Factor 204 |
| 12.2.7 | Concluding Remarks 206 |

13   Nanogas Emulsions as a Displacement Agent *207*
13.1   Oil Displacement by a Newtonian Gasified Fluid *207*
13.2   Oil Displacement by a Non-Newtonian Gasified Fluid *208*
13.3   Mechanism of Observed Phenomena *210*
13.4   Field Application *213*

Nomenclature *217*
References *219*

Part E   Novel Perspective Nanocolloids *225*

14   Metal String Complex Micro and Nano Fluids *227*
14.1   What are Metal String Complexes? *227*
14.2   Thermal Conductivity Enhancement of Microfluids with $Ni_3(\mu3\text{-ppza})_4Cl_2$ Metal String Complex Particles *228*
14.2.1 Microparticles of MSC $Ni_3(\mu3\text{-ppza})_4Cl_2$ *229*
14.2.2 Ni_3 Microfluid *230*
14.2.3 Fluid Stability *230*
14.2.4 Thermal Conductivity *232*
14.2.5 Rheology *235*
14.2.6 Surface Tension *235*
14.2.7 Freezing Points *236*
14.2.8 Concluding Remarks *237*
14.3   Thermophysical Properties of Nano- and Microfluids with $Ni_5(\mu5\text{-pppmda})_4Cl_2$ Metal String Complex Particles *237*
14.3.1 Microparticles of the Metal String Complex $Ni_5(\mu5\text{-pppmda})_4Cl_2$ *238*
14.3.2 Micro- and Nanofluid Preparations *239*
14.3.3 Fluid Stability *240*
14.3.4 Thermal Conductivity *242*
14.3.5 Rheology *245*
14.3.6 Surface Tension *245*
14.3.7 Freezing Points *249*
14.3.8 Concluding Remarks *249*

Nomenclature *251*
References *253*

Appendix A   Determination of Dispersed-Phase Particle Interaction Influence on the Rheological Behavior *259*
Appendix B   Determination of Inflection Points *263*
References *267*
Index *269*

Acknowledgments

The authors would like to recognize the contribution in experimental studies of the following people: Dr. Elhan M. Abbasov, Dr. Hakim F. Abbasov, Dr. Oleg A. Dyshin, and Dr. Rayyet H. Ismayilov.

The authors also thank their families for the support, patience, and understanding they have shown during the preparation of this book.

Introduction

In contempt of growing investments in renewable energy, the oil industry still stays the main source of energy in the world. It is well known that the greatest part of oil resources is still marked as unrecoverable due to the limitation of conventional oil recovery methods. The number of explored oil fields already overtake the fields to be explored. In this regard, the increment in oil production of mature oil fields is very crucial. The oil industry today is standing in the frontier of new pioneering achievements. With a high probability it can be argued that these achievements will be made and have already been committed to laboratories. An oil price decrease has already reduced economic benefits of enhanced oil recovery (EOR) methods due to CAPEX increments. The same challenges are lying ahead of all upstream technologies, such as completion, workover, sand control, etc. The application of new nano-based materials could open new opportunities and find new decisions for conventional problems.

This book aims to cover a theoretical and practical background related to nanocolloid application experience gained over the last decades. Nanocolloids are admitted by the majority of researchers as a new perspective and a promising investigation topic. The high surface area of dispersion phase causes a more reactive behavior compared to conventional counterparts and significantly changes the properties of colloid systems. For instance, the propagation ability of nanogels considerably increases an opening of new opportunities for in-depth fluid diversion techniques as well as gel durability in reservoir conditions. The book consists of five parts divided into chapters to make the experience of reading the book more reader-friendly. The paragraphs that follow briefly describe each part and highlight the main discussed topics.

Part A. Nanocolloids – An Overview consists of two chapters devoted to a brief introduction and classification of colloid systems. It was presented and explained the term "nanocolloid." The chapters describe the main properties of nanocolloids crucial for practical applications in petroleum engineering: for example, stability, rheological behavior, surface tension, and wettability.

Part B. Reservoir Development consists of three chapters devoted to nanocolloid applications in reservoir engineering. The chapters describe reservoir conditions necessary for nanocolloid formation. Nanogas emulsion hydrodynamics at reservoir conditions have been described in detail. Field validation results of the proposed kinetic mechanisms accompanied by technical recommendations for successful implementation were also presented.

Part C. Production Operations consists of four chapters devoted to nanocolloid applications in production operations. The chapters describe the mechanism of the nanoscale dispersion phase impact on physical properties of conventional substances utilized in upstream processes. Particularly, the mechanism of Portland cement reinforcement in the presence of nanoparticles was described and verified. Nanogas emulsions were investigated in terms of sand control applications based on fluidization phenomena. The vibrowave stimulation impact on nano-gas emulsion flow was also reported.

Part D. Enhanced Oil Recovery consists of four chapters devoted to nanocolloid applications for EOR. The chapters describe the impact of nanoparticles on conventional displacement agents. For instance, the addition of nanoparticles significantly changes surface and rheological behavior in surfactant aqueous solutions. The mechanism of the observed phenomena has been explained and specified. Deep flow diversion agents demonstrate improved stability and enhanced physical properties in the existence of nanoparticles in the dispersion phase. Enhanced flow behavior of nanogels improves conformance control due to deeper in-situ propagation and increased thermal stability. Field application results encourage stating nanogas emulsions as effective and profitable displacement agents for oil recovery. Explained mechanisms and reported data referring to observed phenomena could be a good theoretical and practical basis for future investigations.

Part E. Novel Perspective Nanocolloids describes new perspective materials for petroleum engineering. Metal string complexes due to the existence of extra metal ions and paddlewheel geometry have large varieties of metal–metal bonds that lead to significant changes in physical properties. Colloid stability and thermal conductivity have been sustainably improved in the presence of metal string complexes. The overall reported results allow these compounds to be shown as a promising area for future studies and applications in the petroleum industry.

The book is based on materials from many sources, including academic papers, publications from the indexed petroleum journals, academic institutions laboratory report data, as well as the experience of Service and National Oil Companies gained over the last three decades. To make this book a useful and effective guide into the nanocolloids, applied and theoretical basis of the detailed observed results were provided.

Part A

Nanocolloids – An Overview

1

Nanocolloid Classification

1.1 What is a Colloid?

A colloid (colloidal system or mixture) is a heterogeneous matter in which one phase plays a host role (dispersion media) and another phase is present as a stable distinguishable dispersed (dispersion) entity (we can also say particles if the dispersion is from a solid phase). The dispersion phase can have sizes between 1 and 1000 nm [1]. The shape can vary over a very broad range. It is assumed that a colloid should be thermodynamically stable and that the dispersed phase (entities) remains evenly distributed throughout the colloid.

If the dispersed entities are too small, it would not be possible to define them as a phase. At the other extreme, if they are too big then the system would not be stable.

In fact, two common methods exist to determine whether a mixture is a colloid or not:

1) The Tyndall effect. This is based on light scattering by particles in a colloid or a very fine suspension. The light passes easily through a true solution but in colloids the dispersed phase scatters it in all directions, making it hardly transparent.
2) Filtration of the colloid through a semipermeable membrane. In fact, the dispersion phase cannot pass through the membrane and is filtered out of a dispersion medium.

There are three main classifications of colloids according to different properties of the dispersed phase and medium [2, 3].

1.1.1 Colloid Classification

1) **Classification based on the physical state of the dispersion medium and of the dispersed phase.** Using these criteria, colloids can be divided as shown in Table 1.1.
2) **Classification based on the nature of the interaction between the dispersion medium and the dispersed phase.** Using this criterion, colloids can be classified as either lyophilic or lyophobic.
   - Lyophilic (**intrinsic colloids**). A high force of attraction exists between a dispersed phase and the dispersion medium. These types of colloids are very stable and do not require special mixing requirements. Examples are starch, rubber, protein, etc.

Table 1.1 The main types of nanocolloid.

Dispersed phase	Dispersed medium	Name of colloidal solution	Common examples
Solid	Gas	Solid aerosol	Smoke, dust
Solid	Liquid	Sol, gel, suspension	Paints, inks, jellies
Solid	Solid	Solid sol	Alloys, opals
Liquid	Gas	Liquid aerosol	Dust, smog, clouds
Liquid	Liquid	Emulsion	Butter, cream, milk
Liquid	Solid	Solid emulsion (gel)	Butter, cheese, curd
Gas	Liquid	Gas emulsion, foam	Whipped cream, suds
Gas	Solid	Solid foam	Cake, marshmallow, lava

- Lyophobic (**extrinsic colloids**). A very weak force of attraction exists between the dispersed phase and the dispersion medium. These types of colloids are unstable without the application of stabilizing agents and require special mixing procedures. Examples are sols of metals like silver and gold.

3) **Classification based on the types of particles of the dispersed phase colloids.** Using these criteria, colloids can be divided into:

- **Multimolecular colloids.** These are formed by a colloidal range aggregation of atoms or small molecules that have a particle size less than the colloidal range (i.e. with a diameter less than 1 nm) in a dispersion medium. For example, gold consists of various sized particles composed of several atoms of gold.
- **Macromolecular colloids.** These are composed of macromolecules having strong chemical bonds between molecules and of a size in the colloidal range. They are very stable. Examples are starch, proteins, cellulose, etc.
- **Associated colloids.** These are substances that behave as electrolytes at low concentrations, but show colloidal properties at high concentrations due to a micelle formation.

1.1.2 Colloid Evaluation

Colloids are usually evaluated by the following dispersion characteristics [4]:

- **Particle aggregate.** An inter-particle force causes the aggregation of the dispersion phase in small species, where two types of aggregates occur.:
  a) **Nanoparticle agglomeration in dry powder form.** It is very hard to segregate them, even with ultrasonication.
  b) **Nanoparticles in colloids form a large cluster.** This type of cluster requires more than one procedure to disperse them.

Table 1.2 Stability of suspensions with relation to the zeta potential [5].

Zeta potential This is an electrokinetic value for a dispersed phase. The potential is strongly related to the colloid stability.	Average zeta potential in mV
Stability characteristics	
Maximum agglomeration and precipitation	0 to +3
Range of strong agglomeration and precipitation	+5 to −5
Threshold of agglomeration	−10 to −15
Threshold of delicate dispersion	−16 to −30
Moderate stability	−31 to −40
Fairly good stability	−41 to −60
Very good stability	−61 to −80
Extremely good stability	−81 to −100

- **Particle structure (size and shape).** Dispersion phase particle dimensions and shapes have a huge impact on colloids properties. Example are thermophysical properties, rheological behavior, etc.
- **Polydispersity.** In general, a colloid would have various sized particles acting as the dispersion phase. A polydispersity index is a measure of the particle size variation and ranges from 0 to 1. According to international standards, the value of the index below 0.05 is characteristic of monodispersed distribution (all particles have the same size), while the index values above 0.7 indicate a broadly polydisperse system.
- **Zeta potential.** This is an electrokinetic value for a dispersed phase. The potential is strongly related to the colloid stability (see Table 1.2).

1.2 What is a Nanocolloid?

The term nanocolloid is relatively new and considers a colloid that has nanoparticles as a dispersion phase [4, 6]. Nanocolloids have the following main features:

- The dispersion phase is established by compounds in the amorphous or crystalline state, either organic or inorganic, and may demonstrate collective behavior and consists of particles in the 1–100 nm range.
- Repulsion forces act as the main force to prevent a macroscopic phase separation.

Meanwhile it should be mentioned that the dispersion medium is also a very critical factor as well as an interfacial layer covering the dispersion medium. The stability of the obtained colloids depends strongly on the factors mentioned above. This will be discussed and illustrated further later in the book.

Table 1.3 shows the main types of nanocolloid often found in petroleum engineering.

Table 1.3 The main types of nanocolloid often found in petroleum engineering.

Dispersed phase	Dispersed medium	Name of colloidal solution	Applications in petroleum engineering
Solid	Liquid	Nanogel, nanosuspension (nanofluid)	[6–49]
Solid	Solid	Nanosol	[12, 50, 51]
Liquid	Gas	Nanoaerosol	[12]
Liquid	Liquid	Nanoemulsion	[6, 12, 52–54]
Gas	Liquid	Nano gas emulsion, nanofoam	[6]

2

Nanocolloid Properties

2.1 Different Kinds of Interactions in Nanocolloids

As mentioned earlier, the aggregation of nano-sized particles is the main challenge in a discussion of colloid stability. In particular, flocculation is most often observed as a kind of aggregation. In this regard, domination of attractive forces is defined as undesirable, while repulsion is found to be positive for colloid stability. However, collective properties of nanoparticles that are not presented in individual particles should also be taken into account [4].

2.1.1 Van der Waals Interactions

Van der Waals forces are intermolecular forces that generally demonstrate attractive behavior. In some cases, however, they could display repulsive behavior between dissimilar materials in a third medium. In hydrocarbon liquids (i.e. inert polar), uncoated nanoparticles favor the formation of aggregates.

The Van der Waals interaction potential between two particles can be calculated using the Hamaker theory simplified by the Derjaguin approximation:

$$U_{VdW} = -\frac{A}{6}\left[\frac{2R_1R_1}{c^2-(R_1+R_2)^2} + \frac{2R_1R_2}{c^2-(R_1-R_2)^2} + \ln\frac{c^2-(R_1+R_2)^2}{c^2-(R_1-R_2)^2}\right]$$

where $c = R_1 + R_2 + h$ and is the center-to-center distance, h is the distance of separation, and A is the Hamaker constant. Particles that have a high Hamaker constant value show a stronger aggregation behavior in comparison with particles that have a lower value of Hamaker constant.

2.1.2 Electrostatic Interaction

Electrostatic interactions are the attractive or repulsive interactions between charged molecules and particles. There are various mechanisms defining the surface charging in liquids by adsorption, ionization, etc. [55].

To calculate the electrostatic interaction potential, either Derjaguin approximation (DA) or linear-superposition approximation (LSA) is usually used. The environmental conditions should be taken into account for every case as the solution pH, electrolyte media, and ionic strength have a strong impact on these types of interaction.

Nanocolloids for Petroleum Engineering: Fundamentals and Practices, First Edition.
Baghir A. Suleimanov, Elchin F. Veliyev, and Vladimir Vishnyakov.
© 2022 John Wiley & Sons Ltd. Published 2022 by John Wiley & Sons Ltd.

2.1.3 Elastic–Steric Interaction

Steric stabilization is the process by which adsorbed nonionic surfactants or polymers produce strong repulsion between particles in a dispersion [56]. Based on this mechanism, polymer chains become adsorbed onto the surface of colloidal particles in order to avoid the flocculation driven by Van der Waals forces.

2.1.4 Hydrophobic Interaction

Hydrophobic interactions in colloids are defined as interactions that occur between water and low water-soluble molecules (i.e. hydrophobes). This process is mainly associated with an entropy decrease in the liquid due to solubilization of nonpolar molecules. However, it should be noted that hydrophobic interactions are strongly dependent on electrostatic forces (i.e. the presence of ions, charge of the interface, etc.) [55].

2.1.5 Solvation Interaction

The process of attraction and association of solvent molecules with solute molecules/ions involves bonding and Van der Waals forces. In the case of water the process is called hydration. Solvation forces depend on physicochemical properties, such as the wettability angle, surface morphology, etc. [57].

2.1.6 Depletion Interaction

Large particles suspended in dilute solutions of smaller solutes generate an attractive force called the depletion force [58]. For example, these are solutions that contain nonadsorbing polymers. The depletion potential can be calculated by the following formula:

$$U_D = 2\pi R \Pi \left(L_D - \frac{h}{2}\right)^2 \left(1 + \frac{2L_D}{3R} + \frac{h}{6R}\right)$$

where P is the osmotic pressure and L_D is the depletion layer thickness (roughly twice the radius of gyration).

2.1.7 Magnetic Dipole–Dipole Interaction

Magnetic dipole–dipole interaction is observed as a direct interaction between two magnetic dipoles (e.g. also called dipolar coupling). The following formula expresses the dipole – dipole interaction potential:

$$V_M = \frac{8\pi\mu_0 M^2 a^3}{9\left(\frac{s}{a} + 2\right)^3}$$

where M is the magnetization of the material, a the radius of the particles, s the distance between the surfaces of two interacting particles, and μ_0 is the permeability of the vacuum [59].

2.1.8 Osmotic Repulsion

Osmotic repulsion is based on the local increase of osmotic pressure due to a rise in the polymer concentration while overlapping of polymer ligands in colloids takes place [60].

2.2 The Stability of Nanocolloids

As described previously, the stability of a colloid strongly depends on interparticle interactions. To characterize interaction potential between particles and evaluate an aggregation tendency, the Derjaguin–Landau–Verwey–Overbeek (DLVO) theory is the most widely used [61].

The theory states that colloid stability is defined by the total potential energy of the particles (V_T), which is the sum of the attractive (i.e. Van der Waals) V_A and repulsive (i.e. electrostatic) V_R potentials:

$$V_T = V_A + V_R$$

The expression for the attractive potential energy V_A is

$$V_A = -Ar/(12x)$$

where A is the Hamaker constant, r is the radius of the particles (applicable for the spherical particle or approximation), and x is the distance between the surfaces.

The expression for the repulsive potential energy is

$$V_R = 2\pi\varepsilon\varepsilon_0 r\zeta^2 e^{-kx}$$

where ε is the dielectric constant of the solvent, ε_0 is the vacuum permittivity, ζ is the zeta potential, and k is a function of the ionic concentration (while k^{-1} is the characteristic length of the electric double layer, or EDL).

However, some assumptions of DLVO theory that have been found to be critical for nanocolloids should be noted and are presented below [62, 63]:

1) All particles considered to be similar in terms of surface and shape morphology are described as spherical. However, in the case of nano particles this it is not correct.
2) The dispersion mediums inside and outside of a particle are considered to be homogeneous. However, this does not apply in the case of the interfacial layer. For the nano-sized dispersion phase, the interfacial layer could be comparable and even greater than the particle diameter, which could have a great impact on the agglomeration properties of the nano particles.
3) The repulsion and attractive forces are presumed to be independent forces. However, on nanoscale dimensions, this is not an accurate assumption. For example, polarizability increases the coupling between different forces.
4) Electrostatic properties are distributed uniformly on the dispersion phase surfaces.
5) Brownian motion and electrostatic forces govern the distribution of ions.

Furthermore, Suleimanov et al. had shown that, apart from the potential energy of total particles, the gravity and buoyancy forces should be taken into consideration [22].

The authors demonstrated that particle aggregation sizes and consequently the colloid stability strongly depend on equilibrium between the mentioned forces.

The value of dispersion phase particle density at which colloid stabilization is achieved can be determined from the condition of buoyancy when the gravity acting on the particles aggregation is balanced by the buoyancy:

$$N \cdot \frac{4}{3}\pi(r_p + H)^3 \rho_p g = \frac{4}{3}\pi R_a^3 \rho_{bf} g$$

where g is gravitational acceleration, N is the average number of particles in the assembly, r_p is the average radius of the microparticles, H is the thickness of the associated surrounding liquid layer, ρ_p is the density of the dispersion phase, R_a is the average radius of the assembly, and ρ_{bf} is the density of the dispersion medium.

The density of microparticles is calculated as follows:

$$\rho_p = \frac{\rho_{bf}}{N}\left(\frac{R_a}{r_p + H}\right)^3$$

This shows that the particle size distribution and morphology of particles should be considered in terms of nanocolloid stabilization. It is important to note that the Particle Size Distribution Analyzer and Scanning Electron Microscope should be used to obtain the parameters mentioned in the equations.

2.3 Rheology of Nanocolloids

In nanocolloids with a liquid dispersion medium, where a rise in nanoparticle concentration leads to switching in rheology from Newtonian to non-Newtonian (e.g. nanofluids or nanosuspension, gas emulsions, emulsions) behavior, Suleimanov et al. reported the flow characteristic alteration of a surfactant aqueous solution from a Newtonian to a non-Newtonian state in the presence of non-ferrous metal nanoparticles [49]. However, the mechanism behind that deviation has still not been sufficiently studied. The following subsection discusses the mechanism of nanoparticle interaction and the impact of migrations on the rheology of nanocolloids.

2.3.1 Effect of Nanoparticle Interaction on the Colloid Rheology

Nonlinear effects observed in disperse systems with time-dependent rheological properties (e.g. self-oscillations and random oscillations of rheological characteristics) could be explained by specific features of the interaction between their structural elements. The interaction of dispersed phase particles in purely viscous non-Newtonian systems does not notably affect the dynamic behavior of rheological characteristics but rather manifests itself in their steady-state values. The reasons are the relatively small size (~10 µm) of the dispersed phase particles and the absence of any developed three-dimensional structure. The specific features of the rheological behavior of purely viscous non-Newtonian systems can be explained by using the data on their optical characteristics measured in the process of flow.

2.3 Rheology of Nanocolloids

The investigations could be conducted on an experimental setup similar to those used in references [64] and [65]. Light is passed through a cylindrical layer (cell) of the solution in question. The solution flows at a constant speed, e.g. the shear rate. The intensity of the transmitted light, registered as a voltage from a photomultiplier, is recorded. This allows the time dependence of the solution transparency to be observed. The transparency changes when aggregates of the dispersed phase particles are formed or destroyed. Therefore, it is possible to estimate the dynamics of the interaction between the dispersed phase particles by measuring the system transparency.

Example The solution of a 2 wt.% partially hydrolyzed polyacrylonitrile (PAN) was investigated. Its flow dependence at 20 °C is shown in Figure 2.1. It is apparent from the figure that the flow has an initial portion of pseudoplastic character, which is replaced by a dilatant flow (e.g. an increase of viscosity with the shear rate) after the threshold rate (at around $\gamma = 600\,\text{s}^{-1}$) is exceeded.

The flow cell transparency measurements were conducted at four shear rates: in the region of pseudoplastic flow ($\gamma = 400\,\text{s}^{-1}$), at the inflection point of the flow curve ($\gamma = 600\,\text{s}^{-1}$), near the inflection point in the region of the dilatant flow ($\gamma = 800\,\text{s}^{-1}$), and in the region of the developed dilatant flow ($\gamma = 1200\,\text{s}^{-1}$). A Fourier spectral analysis was subsequently performed for these measurements and the results are presented as the power spectral density.

In the region of the pseudoplastic flow, the transparency did not change with time, i.e. the flow of the dispersion was steady. At the inflection point, the steady flow is disturbed and there are periodic transparency oscillations (see Figure 2.2). Further in the transitional region to the dilatant flow, one can observe complex periodic variations in the transparency, with three characteristic frequencies. In the region of the developed dilatant flow, the transparency varies in a quasi-random manner, as indicated by the presence of many harmonics in the power spectral density graph (Figure 2.2c).

On the data basis provided, one could arrive at the following conclusion: when the steady flow of the dispersion is disrupted, the hydraulic resistance increases and, accordingly, the

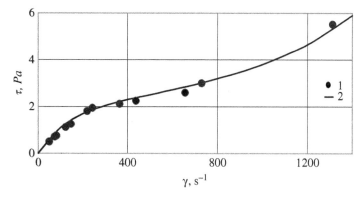

Figure 2.1 Dispersion flow curve. 1 – experimental data points; 2 – polynomial data fitting.

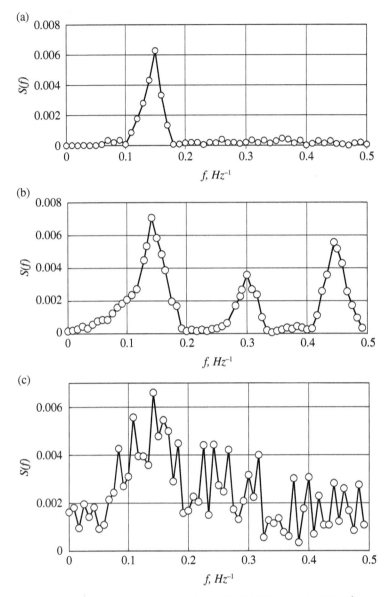

Figure 2.2 Power spectral density. (a) $\gamma = 600$, (b) 800, and (c) 1200 s^{-1}.

apparent viscosity of the system also increases. Each new level of interaction between the structural elements of the system is characterized by a new value of its apparent viscosity.

The majority of researchers consider the observed phenomenon to be quiet rare. According to references [66] and [67], the dilatancy is observed in systems with a high concentration of the solid phase and in coarse dispersions. Dilatant flow for such systems is explained on the basis of the theory of an "excluded-volume." It is assumed that the effective concentration of the solvent in the process of flow constantly decreases due to its higher mobility, the dry friction between dispersed phase particles increases as a result, and, hence, the apparent viscosity of the system increases.

In reference [68] it was shown that strong dilatancy may be observed in a colloid with a dispersion phase smaller than 5 µm. It is explained by preferential particle migration to the walls of the capillary, which leads to an increase in the apparent viscosity of the system.

In reference [69], a dilatant flow was observed in a polymer solution of polymethacrylic acid. The flow behavior was explained by the mechanism involving unfolding of the macromolecule chains with an increase in shear rate and an increase in particle interaction. In some colloids the transition from a low to a high shear rate was associated with the transition from a pseudoplastic flow to a dilatant (e.g. for polyvinyl chloride, hydrolyzed PAN).

The steady pseudoplastic flow of the dispersed phase is not disrupted when the concentration of the partially hydrolyzed PAN in the aqueous solution is below 1 wt.%. Therefore, the disruption of the steady flow is initiated by intensification of inter-particle interaction, which is usually observed with increasing concentration of the dispersed phase and manifests in the formation of dispersed phase particle agglomerates. To verify this assumption, the structure of the dispersion at various polymer concentrations has been investigated using an optical microscope. At low concentrations (up to 0.05 wt.%), the dispersed phase consists of separate deformable randomly oriented elongated (rod-shaped) forms with a size of about 100 nm. As the concentration increases to and above 1 wt.%, the agglomerates (mostly circular or oval) with sizes of 5–10 µm are observed (Figure 2.3). In addition, the average transparency decreases with an increasing shear rate.

In order to determine the uniformity of particle distribution from microphotographs, the dependences of the particle number n in a circle with arbitrary selected radius r, $n(r)$, was measured. Based on the obtained function, the fractal dimensionality of the system's geometrical structure was determined in line with reference [70]. Figure 2.4 shows the fractal

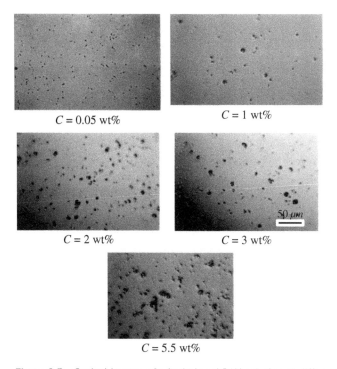

Figure 2.3 Optical images of a hydrolyzed PAN solution at different polymer concentrations.

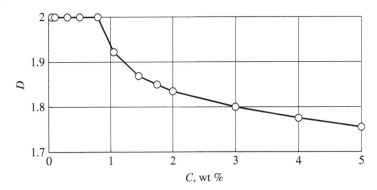

Figure 2.4 Fractal dimension of associates as a function of the polymer concentration C in solution.

dimensionality D function on the polymer concentration. As can be seen from the figure, at concentrations up to 1% the geometrical structure of the system is homogeneous and the fractal dimensionality corresponds to the Euclidean dimensionality of the area. A further increase in the concentration leads to a decrease in the fractal dimensionality of the inhomogeneities, but at the 2% concentration it insignificantly (by 8%) differs from the Euclidean dimensionality of the surface. At low polymer concentrations it is possible in the first approximation to consider the spatial distribution of the dispersed phase as homogeneous.

It is clear that the observed changes in the transparency of sheared dispersion are due to processes of the associated destruction and formation. The absence of a steady state for these processes is evidently attributed to the specificity of the effect of single particles and associates on the hydrodynamic stresses that develop in the dispersion when undergoing shear flow in the region of dilatancy. In order to build a kinetic model that would explain the observed dispersion transparency oscillations, the nature of the dilatant behavior should be studied.

The processes in concentrated dispersed systems mainly show dynamic behavior that leads to a situation when the theoretical study, with accounting for coagulation effects even for the simplest cases, becomes mathematically intricate. The construction of simplified models that provide sufficient accuracy for practical application is very critical.

In references [71–79], based on the kinetic approach, thixotropic processes in dispersed systems were investigated. It was shown that it is possible to predict the rheological properties of a dispersed solution with acceptable accuracy. However, the authors did not consider motion dynamics of the disperse medium and the influence of dispersed-phase particle interaction on the rheological behavior. See Appendix 1 for an illustration of a simple model that takes into account the above-mentioned issues.

2.3.2 Effect of Nanoparticle Migration on the Colloid Rheology

Experiments on various disperse systems have proven that rheological flow dependences cannot be universally described by the exponential law. Indeed, in the case of flow through tubes with variable cross-sections, the rheological function does not obey the exponential law. Moreover, the function of the apparent viscosity upon the flow rate has a nonmonotonic character. Similar effects are also observed in tubes with variable cross-sections for non-Newtonian oils.

However, field data on oil transport indicates that these effects can also be observed in tubes with constant diameters, e.g. during the transport of water–oil emulsions. Moreover, anomalous phenomena also occur when suspensions with solid particles are flowing through the tubes. Experimental investigations of non-Newtonian oil rheology as well as suspensions with solid particles were carried out in order to study these problems in detail. Heavy grades of oil with a complicated rheological behavior were obtained from the Gryazevaya Sopka and Busachi deposits for the investigation. A disperse system model consisting of a clay (bentonite) slurry mixed with added quartz sand (density of 2600 kg/m³) in various fractions (with average particle size of 10^{-4} m) was also examined (Table 2.1). The experimental setup shown in Figure 2.5 was used.

Table 2.1 Basic physical characteristics of the systems studied.

Characteristics and units of measurement	Oil sample from the Gryazevaya Sopka deposit	Oil sample from the Busachi deposit	Clay mortar
Density, kg/m³	980	910	1300
Viscosity, 10^{-3} m, Pa s	198	173	40
Mass concentration of high-molecular compounds of oil, %	28	25	–
Volume concentration of quartz sand, %	–	–	30

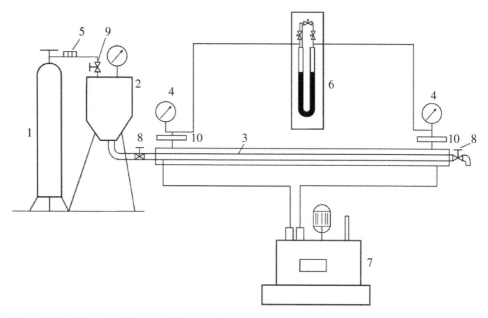

Figure 2.5 Schematic diagram of the experimental setup, where (7) is a high-pressure vessel, (2) an electrical stirrer vessel (for the systems under investigation), (3) a tube of constant diameter (4 or 16×10⁻³ m) with an exchangeable horizontal 2.7 m long section, (4) standard pressure gauges, (5) a gas pressure regulator, (6) a differential pressure gauge, (7) an ultra-thermostat, (8) one-way valves, (9) a locking valve, and (10) separating compensators.

Throughout all the experiments, a constant temperature (287 K) and laminar flow were maintained. The rheological curves or time-dependent flow-rate characteristics were recorded. Figure 2.6 shows a rheological curve for the oil samples from the Busachi deposit, which was used to plot the dependence of the dimensionless apparent viscosity against the value of the shear rate averaged over the cross-section of the tube (Figure 2.7). As can be seen from these figures, the rheological curve is S-shaped, whereas the corresponding dependence of the apparent viscosity upon the shear rate is nonmonotonic. At the inflection point (point A in Figures 2.6 and 2.7), the dilatant behavior becomes pseudo-plastic (in this case, $\gamma \approx 160\ \mathrm{s}^{-1}$). Similar results were obtained for the oil samples taken from the Gryazevaya Sopka deposit.

The dispersed phase of oil systems is known to be composed mainly of asphaltenes, which tend to form supermolecular structures of the size 10^{-6} m. These structures may be deformed under the influence of external forces (for example, shear stress). In accordance with the microrheological study described in reference [80], during the laminar flow of a suspension, solid particles move predominantly near the tube walls, whereas the deformed particles migrate to the center of the flow. Hence, the rheological curves for the oils with a non-Newtonian behavior may be described by the following kinetic mechanism. At a

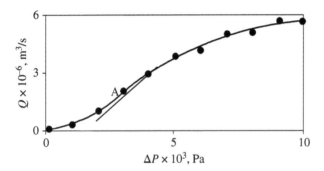

Figure 2.6 Rheological curve for the oil sample from the Busachi deposit, obtained in experiments with a tube 4×10^{-3}m in diameter.

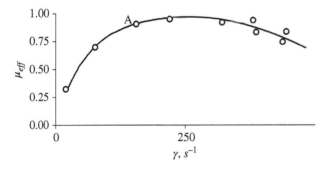

Figure 2.7 Dependence of the dimensionless apparent viscosity on the shear rate averaged over the cross-section of the tube.

relatively low shear rate, the particles of a dispersed phase that are chaotically distributed in a liquid are not deformed (similar to solid particles) and are gradually concentrated (predominantly near the walls of the cylindrical tube), thus resulting in an increase in the apparent viscosity of the system. Then, after attaining a certain threshold shear rate, the particles become deformed, which leads to an additional perturbation of the field of liquid velocities, and they migrate toward the center of the flow. This is reflected by a decrease in the apparent viscosity of the system. In fact, as has been shown above, the character of flow is altered at $\gamma \approx 160 \text{ s}^{-1}$, which is consistent with the results of study [80]. According to it, a dispersed particle 10^{-6} m in size starts to become deformed when the shear rate γ averaged over the cross-section of the tube exceeds 150 s^{-1}.

Considering that for an oil disperse system the densities of the dispersed phase and the dispersion medium are practically equal, the following viscosity function was suggested [81]:

$$h(c) = \frac{1}{\eta(c)} + \frac{A}{r^2} c$$

where $\eta(c)$ is the function of viscosity versus concentration, r is the cylindrical coordinate, c is the concentration of the dispersed phase, and A is a constant. In accordance with reference [81], $A \in [0, \infty]$, $r \in [0, 1]$, and $c \in [0, 1]$. The value of the function $c(r)$ is defined so that $h(c)$ and, accordingly, the flow rate are maximal.

Assuming that the function $\eta(c)$ is expressed by the Einstein equation, we obtain the following expression:

$$h(c) = \frac{1}{1 + ac} + \frac{A}{r^2} c \tag{2.1}$$

By differentiating Equation (2.1) with respect to c, the following expressions can be obtained:

$$h'(c) = -\frac{a}{(1 + ac)^2} + \frac{A}{r^2}$$

$$h''(c) = \frac{2a^2}{(1 + ac)^3} \tag{2.2}$$

An analysis of Equation (2.2) indicates that the function $h(c)$ acquires its maximum value either at $c = 0$ or at $c = 1$. Determining the function $h(c)$ at these values of c,

$$h(1) = \frac{1}{1 + a} + \frac{A}{r^2}, \quad h(0) \equiv 1$$

Since function $h(1)$ decreases with increasing r, $c_m(r)$ for the distribution of the concentration over the cross-section of the tube, thus providing the following maximum flow rate, could be expressed as follows:

$$c_m(r) = \begin{cases} 1, & 0 < r < (1 + a)\sqrt{A/a} \\ 0, & (1 + a)\sqrt{A/a} < r \leq 1 \end{cases} \tag{2.3}$$

For the concentration averaged over the cross-section,

$$F = \int_0^1 c_m(r) r \, dr = \frac{(1+a)^2 A}{2a}$$

i.e. at $A \in (0, a/(1+a)^2)$ and $F \in (0, 1/2)$ the distribution expressed by Equation (2.3) is valid and the system is completely separated.

It should be noted that, in the considered example, when the dispersed phase is moving near the tube wall, complete separation is not obtained. However, according to reference [80], as the particles move from the center of the tube to its walls, an increase in the concentration and, accordingly, in the viscosity takes place.

The next step is an investigation into the impact of the obtained distribution on the liquid flow rate. In this regard, assume that two viscous incompressible liquids with different viscosities are moving in the following manner – one of the liquids is moving in an annulus near the tube walls and the other liquid is moving in the center of the flux. The flow rates of these liquids are determined by the following formula:

$$v_i = -\frac{\Delta P}{4 l \eta_i} r^2 + a_i \ln r + b_i \tag{2.4}$$

where $\Delta P/l$ is the pressure gradient, η_i is the viscosity of a liquid, and a_i and b_i are constants. The liquid near the wall is denoted by the subscript index 1 and the core, the subscript index 2.

Solving Equation (2.4) under the boundary conditions

$$r = R_1, \quad v_1 = 0; \quad r = R_2, \quad \eta_1 \frac{dv_1}{dr} = \eta_2 \frac{dv_2}{dr}$$

and

$$v_1 = v_2, \quad r = 0, \quad \frac{dv_2}{dr} = 0$$

where R_1 is the tube radius and R_2 is the radius of the interface between the two liquids, we obtain the following expression for the total flow rate:

$$Q = \frac{\pi \Delta P R_1^4}{8 \eta_2 l} \left[\varepsilon + (1-\delta)^4 (1-\varepsilon) \right] \tag{2.5}$$

where δ is the thickness of the layer near the wall ($\delta = 1 - R_2/R_1$) and $\varepsilon = \eta_2/\eta_1$.

An analysis of relationship (2.5) indicates that at $\varepsilon > 1$, i.e. when a layer with a lower viscosity moves near the tube walls, the flow rate of the liquid increases with the thickness of the layer near the wall, whereas at $\varepsilon < 1$, the flow rate of the liquid diminishes with the increasing thickness of this layer.

In Figure 2.8, the dependence of the flow rate on the time is represented at a constant pressure drop of 1.1×10^4 Pa for a clay slurry with added quartz sand. It can be seen that the flow rate is a descending function, virtually vanishing over a certain period of time. The scatter of experimental data, which is observed before the steady state is attained, may be attributed to a gradual redistribution of the dispersed phase. Note that a similar scatter is also observed during the flow of rheopectic suspensions.

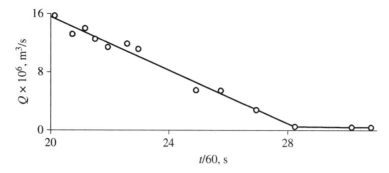

Figure 2.8 Dependence of the clay slurry flow rate with added quartz sand on time at a constant pressure drop of 1.1 × 10⁴Pa, obtained in experiments with a tube 1.6 × 10⁻²m in diameter.

An increase in the input pressure of up to 1.3×10^4Pa results in continued flow of the system. The suspension under investigation possesses a reasonably high sedimentation stability due to the small average size of the particles of the dispersed phase and the relatively high viscosity of the dispersion medium with a comparatively small difference in densities, as well as due to the presence of yield shear stress. This was supported by the fact that the concentration of sand in the outflow of the slurry remains virtually constant.

The dependences of the yield shear stress and the structural viscosity on the concentration for the suspension under consideration have also been studied. It has been established that the yield shear stress and structural viscosity increase rapidly with concentration.

The decrease observed in the flow rate after stopping the flow may be explained as follows. After a certain period of time, in the tube where the suspension has been flowing a distribution of the dispersed phase is established so that its concentration increases with proximity to the tube wall. This, in turn, results in an increase in the yield shear stress of the system.

The next step is to obtain a qualitative conformation of the suggested model. Abbasov found that a density difference between the disperse phase and the dispersion medium cannot be neglected. Then, transforming the function obtained in reference [81] for the instant case, the following expression was achieved:

$$h(c) = -\frac{\rho(c)}{2\eta(c)} + \frac{A}{1-r^2}c$$

where $\rho(c)$ is the function of the density over concentration. For the function $c(r)$, we find a value that results in minimization of the flow rate of the liquid after the function $h(c)$ is maximized. For the function of density versus concentration, this is assumed to be

$$\rho(c) = b + (1-b)c$$

where b is the ratio of the density of the dispersion medium to that of the dispersed phase. The function of viscosity versus concentration is expressed by the formula [82]

$$\eta(c) = a/(a-c), \quad a > 1$$

Then the function $h(c)$ takes the following form:

$$h(c) = \frac{(c-a)[b+(1-b)c]}{2a} + \frac{A}{1-r^2}c$$

Differentiating this with respect to c was obtained as

$$h'(c) = \frac{(2c-a)(1-b)+b}{2a} + \frac{A}{1-r^2}$$

$$h''(c) = \frac{1-b}{a}$$

(2.6)

An analysis of this equation shows that function $h(c)$ attains its maximum at either $c = 0$ or $c = 1$. Determining this function at these values of c was obtained using the following expressions:

$$h(1) = \frac{1-a}{2a} + \frac{A}{1-r^2}, \quad h(0) \equiv -\frac{b}{2}$$

Since function $h(1)$ decreases with r, $[a > b/(1-b)]dr$ was obtained as

$$c_m(r) = \begin{cases} 0, & 0 \le r < \sqrt{1 - \frac{2aA}{a(1-b)-b}} \\ 1, & \sqrt{1 - \frac{2aA}{a(1-b)-b}} < r < 1 \end{cases}$$

(2.7)

The expression for the averaged concentration

$$F = \int_0^1 c_m(r) r\, dr = \frac{aA}{a(1-b)-b}$$

i.e. at $A \in = \frac{a(1-b)-b}{2a}$, $F \in = (0, 1/2)$ the distribution according to Equation (2.7) is observed, and the system is completely separated. Note that, in practice, the complete separation of a liquid does not occur. Therefore, it would be natural to assume that a relatively thin layer near the wall ($\delta = 0.3 - 0.5$) will exhibit a considerably higher concentration of the solid phase and, accordingly, a higher viscosity than those with a uniformly distributed dispersed phase.

In the liquid flow rate evaluation we had assumed that two visco-plastic liquids flow through a cylindrical tube in a circular mode: a viscous-plastic liquid with the structural viscosity η_1 and yield shear stress τ_1 is flowing near the tube walls, whereas another viscous-plastic liquid with η_2 and τ_2 flows in the center of the tube. In this case, $\eta_1 > \eta_2$ and $\tau_1 > \tau_2$. Assuming that the radius of the flow core is relatively small, a simplified approach has been applied, in accordance to which the equation of motion for the viscous-plastic liquid holds throughout the entire range of r values. The flow rate of viscous-plastic liquids is determined from the known formula (using the notations introduced earlier)

$$v_1 = -\frac{\Delta P}{4l\eta_i}r^2 + \frac{\tau_i}{\eta_i}r + a_i \ln r + b_i$$

(2.8)

Solving Equation (2.8) under the following boundary conditions:

$$r = R_1, \quad v_1 = 0; \quad r = R_2, \quad -\tau_1 + \eta_1 \frac{dv_1}{dr} = -\tau_2 + \eta_2 \frac{dv_2}{dr}, \quad v_1 = v_2, r = 0, \quad \frac{dv_2}{dr} = 0,$$

we obtain the following expression for the total flow rate:

$$Q = \frac{\pi \Delta P R_1^4}{8l\eta_2} \left\{ 1 - \frac{8\tau_2 l}{3 \Delta P R_1} \left[(\varepsilon/\varepsilon_1) + D^3 (1 - \varepsilon/\varepsilon_1) \right] - (1 - \varepsilon)(1 - D^4) \right\} \tag{2.9}$$

where $\varepsilon_1 = \tau_2/\tau_1$ and $D = R_2/R_1$.

An analysis of Equation (2.9) indicates that, at $\varepsilon = 1$ and $\varepsilon_1 = 1$, this equation transforms into a simplified Buckingham equation. It should be noted that, in accordance with Equation (2.9), at $\varepsilon_1 \leq \varepsilon$ the yield shear stress of the system proves to be larger than that calculated using the Buckingham equation.

The presented estimates may be interpreted as a qualitative confirmation of the assumption made above. Hence, the results obtained may be explained by the migration of the particles of the dispersed phase toward the tube walls and backwards. This information could be used to control the structure of the flow of disperse systems transported through pipelines.

2.4 Surface Tension. Wettability

2.4.1 Wettability Alteration

Wettability of a porous media plays an important role for oil recovery and directly attributes to surface interactions between solid–fluid or fluid–fluid interfaces [83]. Young's equation expresses the wettability as follows:

$$\cos \theta \sigma_{wo} + \sigma_{sw} = \sigma_{so}$$

where θ is the contact angle, σ_{wo} is the interfacial tension (IFT) on the water–oil interface, σ_{sw} is the IFT on the solid–water interface, and σ_{so} is the IFT on the solid–oil interface.

The interfacial energy defines the interaction behavior between two immiscible phases. In terms of the attraction and repulsive forces concept this means that the attraction between these phases lowers the interfacial energy while the repulsion leads to an interfacial energy increment. The main effect of a wettability alteration is expressed in changes of capillary pressure as well as relative permeability values. In the petroleum industry the surface-active agent's injection is conventionally applied in this regard [84]. However, in the last few decades several studies show that application of nanocolloids could be an effective improvement or even an alternative way of oil recovery.

Wasan and Nikolov, inspired by the impact of nanoparticle intrusion on suspensions spreading and adhesion behavior, studied the spreading behavior of nanofluids containing surfactant micelles [85]. They stated that: "Theoretical investigations have suggested that a solid-like ordering of suspended spheres will occur in the confined three-phase contact region at the edge of the spreading fluid, becoming more disordered and fluid-like towards the bulk phase." This statement had been proved by calculations and dynamic microscopy. The authors have shown that the nanoparticles form crystal-like two-dimensional (2D) layered structures and enhance the spreading dynamic of a micellar fluid at the solid–oil–aqueous contact region. In other words, nanoparticles create an inner contact line that increase the disjoining pressure and the oil–nanofluid interface moves forward, resulting in oil–soil segregation.

Several researches reported that adsorption of nano particles on a sandstone surface leads to wettability alteration. It should be noted that the mechanism of wettability alteration based on DLVO theory (i.e. a disjoining pressure increase) by Ju Binshan et al. from the University of Petroleum, East China, presents a mathematical model of wettability and permeability change caused by adsorption of nanometer structured polysilicon in a porous medium. The model was combined with the study of experiments in the laboratory and field [86].

According to this work, the wettability of the surface of the particle, polysilicon is classified into three types: lipophobic and hydrophilic polysilicon (LHP), neutral wettable polysilicon (NWP), and hydrophobic and lipophilic polysilicon (HLP). Since the wettability of porous walls can be changed by adsorption of nano sized and scaled polysilicon with different wettability, each of the three types may be used in oil development for modifying the flow performance of oil, water, and gas in oil reservoirs.

As an oil-wet reservoir rock can be changed into water-wet rock by adsorption of LHP on porous walls, the relative permeability of the oil phase will increase and the relative permeability of the water phase will decrease. Watercut, in turn, will decline after a water break. Adsorption of NWP on porous walls will eliminate surface tension. Therefore, mobility of the oil phase and displacement efficiency will increase in a water drive reservoir. Thus, LHP and NWP used in oil fields can enhance the oil production rates of oil wells and improve oil recovery. Adsorption of HLP on porous walls will lead to improving the relative amount of the water phase, which can enhance the injection rates for water injection wells. It is necessary to enhance water injection for low permeability reservoirs.

In the work of Lesin et al. [87], the possibility of altering wettability by preferential hydrofilization utilizing a colloidal solution of iron particles is shown. It is noted that hydrofilization brings a rise in oil recovery. The model proposes changing pore surface physicochemical properties by using gas phials formed on the colloidal ferromagnetic particles in a water solution.

Espin et al. proposed a method for altering the wettability of a porous medium via a water-based system with nanoparticles [88]. The developed nanofluid system comprises a combination of organic and inorganic components of nanoparticle size.

The organic nanoparticles are essentially polymeric structures that are adsorbed on the mineral surface, leading to the formation of a film on the mineral surface in order to alter wettability as desired without substantially affecting the formation permeability or porosity.

The inorganic nanoparticles are selected to help control viscosity of the fluid as desired.

It has been found that an intermediate wetting condition is most desirable for enhancing hydrocarbon production from a reservoir.

Aqueous systems are ideal for employment of organic and inorganic components as nanosystems. The nanosystems are well adapted to deposit a thin film of the desired material upon grains of the reservoir in order to modify or alter the wettability to the desired range without significantly impacting upon porosity or permeability. Preferred organic structures include polymeric structures, for example silanes, alkoxysilanes with fluorinated chains, alkylcarbonyl, and the like. These structures advantageously adsorb on to the mineral surface of grains of the reservoir in order to assist in generating the desired thin film. The inorganic components are preferably provided in order to control density and viscosity of the fluid. Essentially, they provide the fluid with a viscosity that is suitable for the desired application. Examples of suitable inorganic nanoparticles include silicon, aluminum, titanium, zirconium, and the like, and combinations thereof.

The water-based fluid system is particularly effective in altering the wettability condition of subterranean reservoirs without substantially reducing reservoir permeability or porosity because the coating system is delivered through nanoparticles carried by the water-based fluid and forms a very thin film around grains of the formation. Nanoparticles for use in the fluid preferably have an average particle size of between about 1 nm and about 200 nm, which combine to provide a versatile system that can be adapted to obtain the desired wetting condition from an existing wetting condition. The systems are also resistant to the harsh conditions experienced in the downhole reservoirs.

2.4.2 Surface Tension

Wettability defines interactions in a fluid–solid interface, particularly adhesive forces. However, one more factor that defines cohesion forces plays an equally important role for enhanced oil recovery – surface tension. Surface tension, the attractive force exerted upon the surface molecules of a liquid by the molecules beneath, tends to draw the surface molecules into the bulk of the liquid and makes the liquid assume the shape with the least surface area (Mirriam-Webster). Surface tension reduction directly increases the capillary number and consequently enhances the oil recovery. In this regard Suleimanov et al. studied the impact of non-ferrous nanoparticle addition on surface tension properties of a surfactant aqueous solution. It was found that the nanoparticles create a thin layer on the surfactant solution interface that increases surface roughness and dramatically decreases the surface tension. Observed phenomena led to an oil recovery increment by 35% in homogeneous and 17% in heterogeneous mediums respectively [49].

There are contradictory study results on the influence of Al_2O_3–H_2O nanofluids on oil recovery [89, 90]. The studies demonstrate minor deviations in surface tension. At the same time some researchers presented surface tension reduction dependencies on volume concentrations and temperatures [91]. Zhu et al. found that surface tension increases with an NP dimension increment [91]. However, it should be noted that there are studies reporting no linear behavior of surface tension in terms of nano-additive concentrations [92].

Nguyen et al. introduced surfactant/polymer inorganic nanocomposites for enhanced oil recovery (EOR) in high temperature reservoirs with high salinity formation water. The SiO_2 nanoparticles in a range of 50–100 nm were added to the polymer solution. The obtained nanocomposite led to IFT reduction and viscosity increments at high concentrations, which shows high thermostability and salt-tolerance. The core flooding experiments at 92 °C on fractured granite showed a 30% reduction in oil saturation and a 6.2% increase in oil recovery.

Li et al. used hydrophilic silica nanoparticles with an average size of 7 nm to investigate potential EOR agents able to mobilize the oil trapped by capillary pressure to increase oil recovery. It was reported that nanofluids reduce IFT at the water–oil interface and lead to wettability alteration. Glass micromodel flooding experiments allowed the release of oil drops trapped in pores to be visualized, while an increase in nanoparticle concentration resulted in oil–water emulsion stabilization. Core flooding experiments showed a 4–5% increment in oil recovery compared to a brine flooding [93].

Ragab et al. published a study comparing the impact of alumina and silica nanoparticles on oil recovery increase. It was reported that IFT reduction and viscosity modification with NP additions as a displacement agent were used in core flood experiments. The authors defined a critical NP concentration as 0.5 wt.% that provides a maximum recovery factor of up to 81%.

Meanwhile, NP addition is very important in terms of a water alternating gas flooding process in intermediate and oil-wet reservoirs. Al Matroushi et al. presented simulation results that show that substitution of water with nanofluid in water alternating gas (WAG) reduced the IFT in an oil–water interface. This results in a wettability alteration from oil-wet to water-wet. Experimental results showed a more than 20% increase in oil recovery compared to the conventional WAG method [94].

Nomenclature

A	Hamaker constant or Hurwitz matrix
C	polymer concentration
D	fractal dimensionality
H	thickness of the associate surrounding the liquid layer
L_D	depletion layer thickness
M	magnetization of the material
N	concentration or average number of particles in the assembly
N_1 and N_2	concentration of elementary inclusions and associates respectively
P	osmotic pressure
ΔP	pressure drop, MPa
R	radius of the tube, m
R_a	average radius of the assembly
R_1	tube radius
R_2	radius of the interface between two liquids
$S(f)$	power spectrum
a	radius of the particles
a_i	constant
a_1 and a_2	nonnegative numbers characterizing the intensity of destruction of elementary inclusions and associates
b	ratio of the density of the dispersion medium to that of the dispersed phase
b_i	constant
c	center-to-center distance or concentration of the dispersed phase
$c_1, c_2,$ and c_3	matrix elements
f	frequency, s^{-1}
g	acceleration of gravity
h	distance of separation
k	function of the ionic concentration
$k(t)$	autocorrelation function
l	length of the tube, m
$n(r)$	integers
q_1	stationary value of N_1

Nanocolloids for Petroleum Engineering: Fundamentals and Practices, First Edition.
Baghir A. Suleimanov, Elchin F. Veliyev, and Vladimir Vishnyakov.
© 2022 John Wiley & Sons Ltd. Published 2022 by John Wiley & Sons Ltd.

q_2	stationary value of N_2
r	coordinate or radius of the particles (note – for the spherical particles) or cylindrical coordinate
r_p	average radius of the microparticles
s	distance between the surfaces of two interacting particles
t	time, s
u	velocity of the liquid at any point of the tube cross-section in the axial direction, m s^{-1}
x	distance between the surfaces
α_1 and α_2	negative numbers characterizing the retardation of the intensity of destruction of particles on increase in their number
β_1 and β_2	nonnegative numbers that determine the rate of recovery of the concentration of particles of both species, s^{-1}
ε	dielectric constant of the solvent
ε_0	vacuum permittivity
ζ	zeta potential
μ_0	permeability of the vacuum
γ	rate of shear, s^{-1}
η	dynamic viscosity of the system, mPa s
η_i	viscosity of a liquid
η_0 and η_1	constants
λ	characteristic numbers
δ	thickness of the layer near the wall
v_1 and v_2	small fluctuations
ρ	density, kg m^{-3}
ρ_p	density of the dispersion phase
ρ_{bf}	density of the dispersion medium
τ	shear stress, Pa
θ	contact angle
σ_{wo}	interfacial tension on the water–oil interface
σ_{sw}	interfacial tension on the solid–water interface
σ_{so}	interfacial tension on the solid–oil interface

References

1 Hiemenz, P.C. and Rajagopalan, R. (ed.) (2016). *Principles of Colloid and Surface Chemistry, Revised and Expanded*. CRC Press.
2 Moore, J.H. and Spencer, N.D. (2019). *Encyclopedia of Chemical Physics and Physical Chemistry*, vol. 3, 2326–2352. CRC Press.
3 Petrucci, R.H., Herring, F.G., Bissonnette, C., and Madura, J.D. (2017). *General Chemistry: Principles and Modern Applications*, 11e. Pearson Canada Inc.
4 Sanchez-Dominguez, M. and Rodriguez-Abreu, C. (ed.) (2016). *Nanocolloids: A Meeting Point for Scientists and Technologists*. Elsevier.
5 Riddick, T.M. (1968). Control of colloid stability through zeta potential. *Blood* 10 (1): 52–68.
6 Huh, C., Daigle, H., Prigiobbe, V., and Prodanovic, M. (2019). *Practical Nanotechnology for Petroleum Engineers*. CRC Press.
7 Al-Olayan A.M. and Alexander-Katz A. New shear thickening dilatancy dispersion based on nano-silica beads for oilfield applications. *SPE-188247-MS. Presented at the Abu Dhabi International Petroleum Exhibition and Conference* held in Abu Dhabi, UAE (13–16 November 2017).
8 Dolog R., Ventura D., Khabashesku V., Darugar Q. Nano-enhanced elastomers for oilfield applications. OTC-27609-MS. *Presented at the Offshore Technology Conference held in Houston*, Texas, USA (1–4 May 2017).
9 Haque M.H., Saini R.K., Sayed M.A. Nano-composite resin coated proppant for hydraulic fracturing. *OTC-29572-MS. Presented at the Offshore Technology Conference held in Houston*, Texas, USA (April 2019).
10 Haroun M., Al Hassan S., Ansari A., et al. Smart nano-EOR process for Abu Dhabi carbonate reservoirs. *SPE-162386. Presented at the Abu Dhabi International Petroleum Exhibition & Conference* held in Abu Dhabi, UAE (11–14 November 2012).
11 Lenchenkov N.S., Slob M., van Dalen E., et al. Oil recovery from outcrop cores with polymeric nano-spheres. SPE-179641-MS. *Presented at the SPE Improved Oil Recovery Conference held in Tulsa*, Oklahoma, USA (11–13 April 2016).
12 Bera, A. and Belhaj, H. (2016). Application of nanotechnology by means of nanoparticles and nanodispersions in oil recovery – a comprehensive review. *Journal of Natural Gas Science and Engineering* 34: 1284–1309.

13 Rong X., Wang Y., Forson K. et al. The influence of nano-silicon material on clay swelling effect. *SPE-196386-MS. Presented at the SPE/IATMI Asia Pacific Oil & Gas Conference and Exhibition held in Bali*, Indonesia (October 2020).

14 Sheshdeh M.J. A review study of wettability alteration methods with regard to nano-materials application. *SPE-173884-MS. Presented at the SPE Bergen One Day Seminar held in Bergen*, Norway (22 April 2015).

15 Shokrlu Y.H., Babadagli T. Transportation and interaction of nano and micro size metal particles injected to improve thermal recovery of heavy-oil. *SPE 146661. Presented at the SPE Annual Technical Conference and Exhibition held in Denver*, Colorado, USA (30 October to 2 November 2011).

16 Soares M.C., Viana M., Oliveira A.L. et al. Improvement of viscosity and stability of polyacrylamide aqueous solution using carbon black as a nano-additive. *OTC 24443. Presented at the Offshore Technology Conference Brasil held in Rio de Janeiro*, Brazil (29–31 October 2013).

17 Skauge T., Hetland S., Spildo K., and Skauge A. Nano-sized particles for EOR. SPE 129933. Presented at the 2010 SPE Improved Oil Recovery Symposium held in Tulsa, Oklahoma, USA (24–28 April 2010).

18 Wei B., Li Q., Li H., et al. Green EOR utilizing well-defined nano-cellulose based nano-fluids from flask to field. SPE-188174-MS. Presented at the Abu Dhabi *International Petroleum Exhibition & Conference held in Abu Dhabi*, UAE (13–16 November 2017).

19 Suleimanov, B.A., Veliyev, E.F., and Dyshin, O.A. (2015). Effect of nanoparticles on the compressive strength of polymer gels used for enhanced oil recovery (EOR). *Petroleum Science and Technology* 33 (10): 1133–1140.

20 Suleimanov B.A., Veliyev E.F., Dyshin O.A. Compressive strength of polymer nanogels used for enhanced oil recovery (EOR). *SPE-181960-MS. Presented at the SPE Russian Petroleum Technology Conference and Exhibition*, Moscow, Russia (24–26 October 2016).

21 Suleimanov B.A. and Veliyev E.F. Nanogels for deep reservoir conformance control. SPE-182534-MS. Presented at the SPE Annual Caspian Technical Conference & Exhibition, Astana, Kazakhstan (1–3 November 2016).

22 Suleimanov, B.A., Ismayilov, R.H., Abbasov, H.F. et al. (2017). Thermophysical properties of nano- and microfluids with [Ni5(μ5-pppmda)4Cl2] metal string complex particles. *Colloids and Surfaces A: Physicochemical and Engineering Aspects* 513: 41–50.

23 Suleimanov, B.A. and Veliyev, E.F. (2017). Novel polymeric nanogel as diversion agent for enhanced oil recovery. *Petroleum Science and Technology* 35 (4): 319–326.

24 Suleimanov, B.A., Veliyev, E.F., and Aliyev, A.A. (2020). Colloidal dispersion nanogels for in-situ fluid diversion. *Journal of Petroleum Science and Engineering, Available online* 19: 107411. In Press.

25 Hammood H.A., Alhuraishawy A.K., Hamied R.S. and AL-Bazzaz W.H. Enhanced oil recovery for carbonate oil reservoir by using nano-surfactant: Part I. *SPE-198662-MS. Presented at the SPE Gas & Oil Technology Showcase and Conference held in Dubai*, UAE (21–23 October 2019).

26 Alhuraishawy A.K., Hamied R.S., et al. Enhanced oil recovery for carbonate oil reservoir by using nano-surfactant: Part II. SPE-198666-MS. *Presented at the SPE Gas & Oil Technology Showcase and Conference held in Dubai*, UAE (21–23 October 2019).

27 Barroso A.L., Marcelino C.P., Leal A.B., et al. New generation nano technology drilling fluids application associated to geomechanic best practices: Field trial record in Bahia – Brazil. *OTC-28731-MS. Presented at the Offshore Technology Conference held in Houston*, Texas, USA (30 April–3 May 2018).
28 Jang, S.P. and Choi, S.U.S. (2007). Effects of various parameters on nanofluid thermal conductivity. *Journal of Heat Transfer* 129: 617–623.
29 Elkady M. Application of nanotechnology in EOR, polymer-nano flooding the nearest future of chemical EOR. SPE-184746-STU. *Presented at the SPE international Student Paper Contest at the SPE Annual Technical Conference and Exhibition held in Dubai*, UAE (26–28 September 2016).
30 Fakoya M.F. and Shah S.N. Rheological properties of surfactant-based and polymeric nanofluids. SPE 163921. *Presented at the SPE/ICoTA Coiled Tubing & Well Intervention Conference & Exhibition held in The Woodlands*, Texas, USA (26–27 March 2013).
31 Hendraningrat, L., Li, S., and Torsæter, O. (2013). A coreflood investigation of nanofluid enhanced oil recovery. *Journal of Petroleum Science and Engineering* 111: 128–138.
32 Jafariesfad N., Gong Y., Geiker M.R., and Skalle P. Nano-sized MgO with engineered expansive property for oil well cement systems. *SPE-180038-MS. Presented at the SPE Bergen One Day Seminar held in Bergen*, Norway (20 April 2016).
33 Kumar D., Chishti S.S., Rai A., and Patwardhan S.D. Scale inhibition using nano-silica particles. SPE 149321. *Presented at the SPE Middle East Health, Safety, Security, and Environment Conference and Exhibition held in Abu Dhabi*, UAE (2–4 April 2012).
34 Kutty Sh. M., Kuliyev M., Hussein M.A., Chitre S. et al. Well performance improvement using complex nano fluids. SPE-177831-MS. *Presented at the Abu Dhabi International Petroleum Exhibition and Conference held in Abu Dhabi*, UAE (9–12 November 2015).
35 Mashat A., Gizzatov A., Abdel-Fattah A. Novel nano-surfactant formulation to overcome key drawbacks in conventional chemical EOR technologies. *SPE-192135-MS. Presented at the SPE Asia Pacific Oil & Gas Conference and Exhibition held in Brisbane*, Australia (23–25 October 2018).
36 Murtaza M., Mahmoud M., Elkatatny S. et al. Experimental investigation of the impact of modified nano clay on the rheology of oil well cement slurry. *IPTC-19456-MS. Presented at the International Petroleum Technology Conference held in Beijing*, China, (26–28 March 2019).
37 Omar A.B., Alawani N., Al Moajil A., and Aldarweesh S. Alcohol ethoxylate nano surfactant: Surface tension and compatibility with acidizing additives. *OTC-28801-MS. Presented at the Offshore Technology Conference held in Houston*, Texas, USA (30 April–3 May 2018).
38 Omotosho Y.A., Falode O.A., Ojo T.I. Experimental investigation of nanoskin formation threshold for nano-enhanced oil recovery nano-EOR. *SPE-198857-MS. Presented at the Nigeria Annual International Conference and Exhibition held in Lagos*, Nigeria (5–7 August 2019).
39 Philip, J., Shima, P.D., and Raj, B. (2008). Nanofluid with tunable thermal properties. *Applied Physics Letters* 92: 043108.
40 Prasher, R., Song, D., and Wang, J. (2006). Measurements of nanofluid viscosity and its implications for thermal applications. *Applied Physics Letters* 89: 133108.
41 Ragab A.M.S., Hannora A.E. An experimental investigation of silica nano particles for enhanced oil recovery applications. SPE-175829-MS. *Presented at the SPE North Africa Technical Conference and Exhibition held in Cairo*, Egypt (14–16 September 2015).

42 Ogolo N.A., Olafuyi O.A., Onyekonwu M.O. Enhanced oil recovery using nanoparticles. SPE-160847-MS. *Presented at the 2012 SPE Saudi Arabia Section Technical Symposium and Exhibition held in AlKhobar*, Saudi Arabia (8–11 April 2012).

43 Taha N.M. and Lee S. Nano graphene application improving drilling fluids performance. IPTC-18539-MS. *Presented at the International Petroleum Technology Conference held in Doha*, Qatar (6–9 December 2015).

44 Tarek M. Investigating nano-fluid mixture effects to enhance oil recovery. *SPE-178739-STU. Presented at the SPE International Student Paper Contest at the SPE Annual Technical Conference and Exhibition held in Houston*, Texas, USA (28–30 September 2015).

45 Zhang J., Li L., Wang Sh., et al. Novel micro and nano particle-based drilling fluids: Pioneering approach to overcome the borehole instability problem in shale formations. SPE-176991-MS. *Presented at the SPE Asia Pacific Unconventional Resources Conference and Exhibition held in Brisbane*, Australia (9–11 November 2015).

46 Zhao T., Chen Y., Pu W., et al. Enhanced oil recovery using a potential nanofluid based on the halloysite nanotube/silica nanocomposites. *SPE-193641-MS. Presented at the SPE International Conference on Oilfield Chemistry held in Galveston*, Texas, USA (March 2019).

47 Wang, L., Chen, H., and Witharana, S. (2013). Rheology of nanofluids: A review. In: *Recent Patents on Nanotechnology*, vol. 7, No. 3, 232–246 (15). Bentham Science Publishers.

48 Mahbubul, I.M. (2019). *Preparation, Characterization, Properties, and Application of Nanofluid*. United Kingdom: Elsevier Inc.

49 Suleimanov, B.A., Ismailov, F.S., and Veliyev, E.F. (2011). Nanofluid for enhanced oil recovery. *Journal of Petroleum Science and Engineering* 78 (2): 431–437.

50 Tabatabaei M., Taleghani A.D., Alem N. Economic nano-additive to improve cement sealing capability. *SPE-195259-MS. Presented at the SPE Western Regional Meeting held in San Jose*, California, USA (23–26 April 2019).

51 Choolaei, M., Rashidi, A.M., Ardjmand, M. et al. (2012). The effect of nanosilica on the physical properties of oil well cement. *Materials Science and Engineering: A* 538: 288–294.

52 Maserati G., Daturi E., Belloni A., Del Gaudio L. et al. Nano-emulsions as cement spacer improve the cleaning of casing bore during cementing operations. *SPE 133033. Presented at the SPE Annual Technical Conference and Exhibition held in Florence*, Italy (19–22 September 2010).

53 Al-Anazi M.S., Al-Khaldi M.H., Fuseni A., and Al-Marshad K.M. Use of nano-emulsion surfactants during hydraulic fracturing treatments. *SPE-171911-MS. Presented at the Abu Dhabi International Petroleum Exhibition and Conference held in Abu Dhabi*, UAE (10–13 November 2014).

54 Mensah A.E., Opeyemi A., Shaibu M. Effects of nano-particles (Al, AL_2O_3, Cu, CuO) in emulsion treatment and separation. *SPE 167508. Presented at the Nigeria Annual International Conference and Exhibition held in Lagos*, Nigeria (30 July–1 August 2013).

55 Chandler, D. (2005). Interfaces and the driving force of hydrophobic assembly. *Nature* 437 (7059): 640–647.

56 Walker, D.A., Kowalczyk, B., de La Cruz, M.O., and Grzybowski, B.A. (2011). Electrostatics at the nanoscale. *Nanoscale* 3 (4): 1316–1344.

57 Wood, J.A. and Rehmann, L. (2014). Geometric effects on non-DLVO forces: Relevance for nanosystems. *Langmuir* 30 (16): 4623–4632.

58 Jenkins, P. and Snowden, M. (1996). Depletion flocculation in colloidal dispersions. *Advances in Colloid and Interface Science* 68: 57–96.

59 De Vicente, J., Delgado, A.V., Plaza, R.C. et al. (2000). Stability of cobalt ferrite colloidal particles. Effect of pH and applied magnetic fields. *Langmuir* 16 (21): 7954–7961.

60 Fritz, G., Schadler, V., Willenbacher, N., and Wagner, N.J. (2002). Electrosteric stabilization of colloidal dispersions. *Langmuir* 18: 6381–6390.

61 Derjaguin, B.V., Churaev, N.V., and Muller, V.M. (1987). The Derjaguin-Landau-Verwey-Overbeek (DLVO) theory of stability of lyophobic colloids. In: *Surface Forces*, 293–310. Boston, MA: Springer.

62 Striolo, A. and Egorov, S.A. (2007). Steric stabilization of spherical colloidal particles: Implicit and explicit solvent. *The Journal of Chemical Physics* 126 (1): 014902.

63 Ninham, B.W. (1999). On progress in forces since the DLVO theory. *Advances in Colloid and Interface Science* 83 (1–3): 1–17.

64 Bibik, E.E. and Lavrov, I.S. (1970). Measurement of the cohesive force of particles in aggregated disperse systems. *Colloid Journal* 32 (4): 483–488.

65 Baran, S. and Gregori, D. (1996). Flocculation of kaolin suspensions by cation polyelectrolytes. *Colloid Journal* 58 (1): 13–18.

66 Makovei, N. (1982). *Hidraulica Forajului*. Editura Tehnică: București.

67 Ur'ev, N.B. and Choi, S.V. (1996). Rheological characteristic of structured dispersions displaying dilatant properties. *Colloid Journal* 58 (6): 862–864.

68 Reynolds, W.W. (1965). *Physical Chemistry of Petroleum Solvents*. New York: Reinhold.

69 Zubov, P.I., Lipatov, Y.S., and Kanevskaya, E.A. (1961). Concerning the dependence of polymer-chain conformation in a solution on solution concentration. *Dokl. Akad. Nauk SSSR* 141 (3): 387–388.

70 Zosimov, V.V. and Tarasov, D.N. (1997). Dynamic fractal structure of emulsions due to motion and interaction of the particles: Numerical simulation. *Journal of Experimental and Theoretical Physics*. 111 (4): 1314–1319.

71 Svalov, A.M. (1987). On a certain model of thixotropic systems. *Colloid Journal* 49 (4): 799–802.

72 Kharin, V.T. (1984). Rheology of viscoelastic thixotropic fluids such as oil and polymer solutions and melts. *Fluid Dynamics* 19: 355–360.

73 Wilkinson, W. (1960). Non-Newtonian fluids. In: *Fluid Mechanics, Mixing and Heat Transfer*. New York: Pergamon Press.

74 Suleimanov, B.A. (1995). On filtration of disperse systems in a nonuniform porous medium. *Colloid Journal* 57 (5): 743–746.

75 Panakhov, G.M. and Suleimanov, B.A. (1995). Characteristic features of the flows of suspensions and oil disperse systems. *Colloid Journal* 57 (3): 386–390.

76 Quemada, D. (1998). Rheological modeling of complex fluids. I. The concept of effective volume friction revisited. *The European Physical Journal Applied Physics* 1: 119–127.

77 Grosberg, A.Y., Khokhlov, A.R., and Pande, V.S. (1994). *Statistical Physics of Macromolecules*. New York: AIP Press.

78 Slipenyuk, T.S. (1998). Influence of polymers on formation of flocculation structures in bentonite clay suspensions. *Colloid Journal* 60 (1): 70–72.

79 Khasanov, M.M. and Yagubov, I.N. (1990). Fluctuations in the rate of flow during filtration of polymer solutions. *Journal of Engineering Physics and Thermophysics* 59 (2): 211–215.

80 Goldsmit G. (1973). *Mekhanika*. no. 6, p. 69.
81 Pavlovskii Yu.N (1967). *Izv. Akad. Nauk SSSR*, no. 2, p. 160.
82 Perry, R.H., Green, D.W., and Maloney, J.O. (1997). *Perry's Chemical Engineers' Handbook*. The McGrawHill Companies, Inc.
83 Bera, A., Mandal, A., and Kumar, T. (2015). The effect of rock-crude oil-fluid interactions on wettability alteration of oil-wet sandstone in the presence of surfactants. *Petroleum Science and Technology* 33 (5): 542–549.
84 Vishnyakov, V., Suleimanov, B., Salmanov, A., and Zeynalov, E. (2019). *Primer on Enhanced Oil Recovery*, 1e. Imprint: Gulf Professional Publishing.
85 Wasan, D.T. and Nikolov, A.D. (2003). Spreading of nanofluids on solids. *Nature* 423 (6936): 156–159.
86 Ju B., Dai S., Luan Z., et al. A study of wettability and permeability change caused by adsorption of nanometer structured polysilicon on the surface of porous media. *SPE-77938-MS*. Presented in the SPE Asia Pacific Oil and Gas Conference and Exhibition, Melbourne, Australia (8–10 October, 2002).
87 Lesin, V.I., Mikhailov, N.N., and Sechina, L.S. (2002). Use of colloidal particle of iron in water for oil and gas reservoir porous media surface modification. *Geology, Geophysics and Oil and Gas Reservoir Engineering* 2.
88 Espin D., Ranson A., Chavez J.C. et al. (2003). Water-based system for altering wettability of porous media. U.S. Patent No. 6,579,572. Washington, DC: U.S. Patent and Trademark Office.
89 Golubovic, M.N., Hettiarachchi, H.M., Worek, W.M., and Minkowycz, W.J. (2009). Nanofluids and critical heat flux, experimental and analytical study. *Applied Thermal Engineering* 29 (7): 1281–1288.
90 Das, S.K., Putra, N., and Roetzel, W. (2003). Pool boiling characteristics of nano-fluids. *International Journal of Heat and mass transfer* 46 (5): 851–862.
91 Zhu, B.J., Zhao, W.L., Li, J.K. et al. (2011). Thermophysical properties of Al_2O_3-water nanofluids. In: *Materials Science Forum*, vol. 688, 266–271. Trans Tech Publications Ltd.
92 Vafaei, S., Purkayastha, A., Jain, A. et al. (2009). The effect of nanoparticles on the liquid–gas surface tension of Bi_2Te_3 nanofluids. *Nanotechnology* 20 (18): 185702.
93 Li S., Hendraningrat L., Torsaeter O. Improved oil recovery by hydrophilic silica nanoparticles suspension: 2-phase flow experimental studies. *IPTC-16707-MS*. Presented at the International Petroleum Technology Conference, Beijing, China (26–28 March 2013).
94 Al Matroushi M., Pourafshary P., Al Wahaibi Y.Possibility of nanofluid/gas alternating injection as an EOR method in an oil field. *SPE-177434-MS*. Presented at the Abu Dhabi International Petroleum Exhibition and Conference, Abu Dhabi, UAE (9–12 November, 2015).

Part B

Reservoir Development

3

Reservoir Conditions for Nanocolloid Formation

3.1 In-Situ Formation of Nanogas Emulsions

During an oil reservoir development, a flow of gasified fluids occurs. Gasified liquids at pressures above the bubble point are usually studied as a homogeneous fluid since the classical theory of phase transitions suggests a supercritical nucleation of new phase particles. However, studies conducted by Mirzajanzade [1] have shown the flow deviation (e.g. particularly that the flow rate increases by two to three times) of gasified Newtonian liquids at a pressure above the bubble point during a stationary flow. At the same time, the liquid demonstrated nonequilibrium properties while in nonstationary conditions. Experimental data were explained by intense subcritical nucleation. The first time such a rheological phenomenon involving a solid–liquid phase transition was found by Ubbelohde [2] and named pre-melting.

The physical reasons and appropriate mechanisms of these new phase particles of supercritical nucleation have been investigated in many contemporary works. In reference [3], Frenkel put forward the theory of heterophase fluctuations. In reference [4], Zeldovich showed that the heterophase fluctuations are high while the surface tension between phases tends to approach a zero value and the "transient phenomena" greatly depend on surface effects.

This assumption has been confirmed in a number of recent works [5, 6] where the subcritical nuclei's stabilization of a new phase is related to the release of SAAs (surface active agents) on a particle's surface. The SAA traces are also found in real systems that have not been specially treated [6]. According to reference [7], the availability of surfactants in liquids does not remove the problem of nuclei stability, as the surfactant admixture can only reduce the dissolution velocity and does not prevent total nuclei disappearance. However, in references [7] and [8] it has been reported that stabilization could be caused by an electric charge on the nucleus surface. It is important to consider the combined effect of surface tension and electric charges.

3.1.1 Stability of the Subcritical Gas Nuclei

To prove the previous statement, in the nucleation process in a gas-saturated liquid a model should be considered in terms of the interfacial tension dependence on the curvature radius and electrical charge that is homogeneously distributed over the surface of the spherical ideal gas bubble.

The energy required for the formation of one gas molecule in a fluid, considering the electrical charge, homogeneously distributed over the nucleus surface, can be found from the equation

$$\Delta u = 4\pi \frac{d(\sigma(r)r_n^2)}{dN} + \frac{d}{dN}\left[\frac{(ze)^2}{4\pi\varepsilon\varepsilon_0 r_n}\right] \qquad (3.1)$$

where r_n is the nucleus radius, N is the number of molecules in one nucleus, ze is the electrical charge on the nucleus surface, e is the elementary charge, ε is the dielectric permittivity, ε_0 is the vacuum permittivity, and $\sigma(r)$ is the surface tension. N, the number of molecules in one nucleus, is defined by

$$N = \frac{4\pi r^3}{3v_g} \qquad (3.2)$$

where v_g is the volume of the gas molecule.

The surface tension of the nucleus expressed by Tolman's equation, ignoring the quadratic terms, is defined as

$$\sigma(r) = \frac{\sigma_0}{1 + \frac{2\delta}{r}}$$

or

$$\sigma(r) = \sigma_0\left(1 - \frac{2\delta}{r}\right) \qquad (3.3)$$

where σ_0 is the surface tension of the plane interface and δ is the thickness of the nucleus surface layer. Formula (3.3) is correct only at $r \gg \delta$. At $r \sim \delta$, the dependence $\sigma = f(r)$ becomes linear:

$$\sigma(r) = k\,r$$

where k_0 is the proportionality coefficient.

Then, at the beginning of nucleation, when the nucleus radius is still small, taking into account Equations (3.2) and (3.3), the formula (3.1) can be expressed as follows:

$$\Delta u = 3k_0 v_g - \frac{v_g(ze)^2}{16\pi^2 \varepsilon\varepsilon_0 r^4} \qquad (3.4)$$

According to Boltzmann's law,

$$P = P_b \exp\left(-\frac{\Delta u}{kT}\right)$$

then

$$\ln\left(\frac{P}{P_b}\right) = -\frac{3k_0 v_g}{KT} + \frac{v_g(ze)^2}{16\pi^2 \varepsilon\varepsilon_0 r^4 KT} \qquad (3.5)$$

where P is the external pressure, P_b is the pressure in the bubble (or saturation pressure), K is Boltzmann's constant, and T is the absolute temperature.

The second term in Equation (3.5) substantially changes the nature of the dependence and is equal to a non-zero value of r at $\ln(P/P_b) = 0$.

At $r \gg \delta$, from Equation (3.1), considering Equations (3.2) and (3.3), the following is obtained:

$$\ln\left(\frac{P}{P_b}\right) = -\frac{2v_g\sigma_0}{KTr}\left(1 - \frac{\delta}{r}\right) + \frac{v_g(ze)^2}{16\pi^2\varepsilon\varepsilon_0 r^4 KT} \quad (3.6)$$

Since $\delta/r \ll 1$, Equation (3.6) can be expressed as

$$\ln\left(\frac{P}{P_b}\right) = -\frac{2v_g\sigma_0}{KTr} + \frac{v_g(ze)^2}{16\pi^2\varepsilon\varepsilon_0 r^4 KT} \quad (3.7)$$

Figure 3.1 shows the dependencies $\ln(P/P_b)$ on (r/δ), calculated from Equation (3.7). It can be seen from Figure 3.1 that the formation of the subcritical nuclei begins above the gas saturation pressure and the main input is associated with the surface electrical charge.

The threshold radius of the stable gas bubble can be found from expression (3.5) at $\ln(P/P_b) = 0$ for the case $r \sim \delta$:

$$r_0 = \sqrt[4]{\frac{(ze)^2}{48k_0\pi^2\varepsilon\varepsilon_0}}$$

At $r \gg \delta$, the threshold radius can be determined from expression (3.7) at $\ln(P/P_b) = 0$:

$$r_0 = \sqrt[3]{\frac{(ze)^2}{32\sigma_0\pi^2\varepsilon\varepsilon_0}}$$

Given the following: $z = 1$, $e = 1.6 \times 10^{-19}$ C, $\varepsilon = 1.5 \div 2.5$, $\varepsilon_0 = 8.85 \times 10^{-12}$ F/m^{-1}, $\sigma = 0.25 \div 25$ mN/m^{-1}, and $k_0 = 2...5 \times 10^5$ N/m^{-2}, it is possible to calculate that the stable

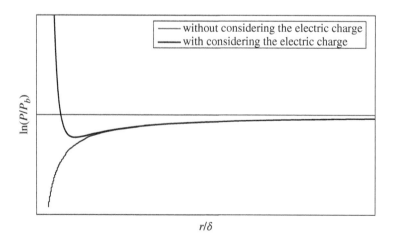

Figure 3.1 Dependence of $\ln(P/P_b)$ on (r/δ).

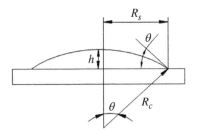

Figure 3.2 Morphology of nanobubbles.

subcritical nuclei radius is in range of 1 to 5 nm. Equations (3.5) and (3.7) show that the presence of an electrical charge on the nucleus surface practically corresponds to the decrease in the surface tension σ and nucleation of the gas phase above the saturation pressure.

The morphology of the subcritical nuclei is of great importance during their stabilization. Study [9] presents the photos of air nanobubbles and their morphology, obtained by the Atomic Force Microscopy. It is shown that the nanobubbles have an irregular flat shape in the form of a pancake; moreover, the height of the bubbles is 20–30 nm and the area of the surface occupied by them is $4 \div 6 \times 10^3$ nm². The nanobubbles are similar to the spherical segment in their shape (see Figure 3.2) [10, 11]. Furthermore, the contact angle between water and gas, the radius of the bubble curvature, and the volume of the spherical segment can be found from the following expressions [10]:

$$\theta = 2tg^{-1}\left(\frac{h}{R_s}\right), R_c = \frac{R_s^2 + h^2}{2h}, V_c = \frac{1}{3}\pi h^2(3R_c - h)$$

The results of the described experimental studies allow for a new look at the stability of nanobubbles. Indeed, the curvature radius of the spherical segment is much higher than in the case when the nanobubble was a hemisphere ($R_s = R_c = h$), since $R_c \approx 25 - 35h$ and $R_c \approx 50h$. The evaluations of the pressure inside the bubble according to the Laplace formula, provided in reference [10], showed that they give 21.5 and 0.45 MPa for the air nanobubbles in the aqueous medium in the case of the morphology of the hemisphere ($R_c = R_s = h = 6.6$ nm) and the spherical segment. Thus, the probability of collapse of the nanobubble with the morphology of the segment is much lower than with the morphology of the hemisphere. At the same time, if the nanobubble is stable at an internal pressure of 21.5 MPa, it would be virtually nondeformable (which is not supported by the experimental data [11]) and the liquid slippage probability would be low.

3.2 In-Situ Formation of Nanoaerosoles

During the oil reservoir development, flow of gas-condensate fluids occurs. A gas-condensate fluid is a hydrocarbon aerosol dispersion that contains gas as the dispersion medium and liquid as the dispersion phase. It has a significant impact on the flow in porous media due to the liquid secretion that occurs when the pressure drops below the dew point pressure. Therefore, it leads to a sharp decrease in the relative gas permeability, ultimately restricting the gas flow.

Thus, the study of hydrocarbon aerosol flows in porous media is of significant interest, especially when subcritical gas-condensate fluids are involved.

3.2.1 Stability of the Subcritical Liquid Nuclei

Several studies reported that surface tension and electric charges formed on nuclei surfaces by selective adsorption of ions has a great impact on the subcritical nuclei stabilization [8, 12, 13]. The energy of one liquid molecule formation during the subcritical nucleation while retrograde condensation in terms of surface and electrical effects are expressed as follows [3, 8, 13]:

$$\Delta u = 4\pi \frac{d(\sigma(r_n) r_n^2)}{dN} + \frac{d}{dN}\left[\frac{(ze)^2}{4\pi \varepsilon_1 \varepsilon_0 r_n}\right] \quad (3.8)$$

where ε_1 is the dielectric permittivity.

The number of molecules in a nucleus is defined by the expression:

$$N = \frac{4\pi r_n^3}{3 V_f} \quad (3.9)$$

where V_f is the molecular volume. The surface tension is calculated by Tolman's equation [14] without the quadratic terms:

$$\sigma(r_n) = \sigma_0 \left(1 - \frac{2\delta}{r_n}\right) \quad (3.10)$$

Equation (3.10) is correct only when $r_n \gg \delta$. When $r_n \sim \delta$,

$$\sigma(r_n) = a_0 r_n \quad (3.11)$$

where a_0 is the proportionality coefficient. Upon inserting Equations (3.9) and (3.11) into Equation (3.8), we get

$$\Delta u = 3 a_0 v_B - \frac{V_f (ze)^2}{16\pi^2 \varepsilon_1 \varepsilon_0 r_n^4} \quad (3.12)$$

If Boltzmann's law is used with Equation (3.12) then

$$\ln(P/P_c) = -\frac{3 a_0 V_f}{KT} + \frac{V_f (ze)^2}{16\pi^2 \varepsilon_1 \varepsilon_0 r_n^4 KT} \quad (3.13)$$

where P_c is the pressure in the nucleus (or dewpoint pressure).

As can be seen from Equation (3.13), r_n is non-zero at $\ln(P/P_c) \geq 0$. That is, there are some stable subcritical liquid nuclei present in the system.

At $r_n \gg \delta$, upon inserting Equations (3.9) and (3.10) into Equation (3.8) and assuming that $(\delta/r_n) \ll 1$, the following equation is obtained:

$$\ln(P/P_c) = -\frac{2 V_f \sigma_0}{KT r_n} + \frac{V_f (ze)^2}{16\pi^2 \varepsilon_1 \varepsilon_0 r_n^4 KT} \quad (3.14)$$

As in the gas case, the surface electrical charge contributes to the formation of stable liquid nuclei above the dew point pressure.

When $r_n \sim \delta$, the radius of the stable subcritical nucleus can be determined from Equation (3.13) at $\ln(P/P_c) = 0$:

$$r_{n0} = \sqrt[4]{\frac{(ze)^2}{48a_0\pi^2\varepsilon_1\varepsilon_0}}$$

The radius of the nucleus at $r_n \gg \delta$ can be determined from Equation (3.14) at $\ln(P/P_c) = 0$:

$$r_{n0} = \sqrt[3]{\frac{(ze)^2}{32\sigma_0\pi^2\varepsilon_1\varepsilon_0}}$$

The above expressions use the following parametric values: $z = 1 \div 2$, $e = 1.6 \times 10^{-19}$ C, $\varepsilon_1 = 1.5 \div 2.5$, $\varepsilon_0 = 8.85 \cdot 10^{-12}$ F/m^{-1}, $\sigma = 0.25 \div 25$ mN/m^{-2}, and $a_0 = 2.5 \times 10^5$ N/m^2. The average radius of the stable subcritical nucleus is 1–10 nm.

4

Nanogas Emulsions in Oil Field Development

4.1 Hydrodynamics of Nanogas Emulsions

4.1.1 Flow Mechanism of Gasified Newtonian Liquids

4.1.1.1 Annular Capillary Flow Scheme

Let us assume that a gas nucleus is formed and then absorbed on a capillary wall surface. This will lead to a situation of mobility modification in the near-wall layer.

In order to determine the flow rate, an annular scheme of capillary flow can be applied. The velocity of the steady-state fluid flow in a capillary is determined by the following equation:

$$v_i = -\frac{\Delta P}{4l\eta_i} r^2 + a_{1i} \ln r + a_{2i} \qquad (4.1)$$

where $\Delta P/l$ is the pressure gradient and η_i is the fluid viscosity (the fluids in the central part of the capillary and in the near-wall layer are denoted by subscripts 1 and 2 respectively). Constants a_{1i} and a_{2i} are found from the following boundary conditions:

$$v_2 = 0; r = R$$

$$\eta_2 \frac{dv_2}{dr} = \eta_1 \frac{dv_1}{dr}, v_1 = v_2, r = R_0$$

$$\frac{dv_1}{dr} = 0, r = 0$$

where R is the capillary radius and R_0 is the radius of the contact line.

In order to estimate the parameters of Equation (4.1) from the boundary conditions, the following expressions for the velocity were derived:

$$v_1 = \frac{\Delta P R^2}{4l\eta_1} \left[\left(\frac{R_0}{R}\right)^2 - \left(\frac{r}{R}\right)^2 + \varepsilon\left(1 - \frac{R_0^2}{R^2}\right) \right] \qquad (4.2)$$

and the relative liquid flow rate is

$$Q_1 = \frac{Q}{Q_0} = S^4 \left[1 + \frac{2\varepsilon}{S^2}(1 - S^2) \right]$$

Nanocolloids for Petroleum Engineering: Fundamentals and Practices, First Edition.
Baghir A. Suleimanov, Elchin F. Veliyev, and Vladimir Vishnyakov.
© 2022 John Wiley & Sons Ltd. Published 2022 by John Wiley & Sons Ltd.

or

$$Q_1 = (1-\xi)^4 \left[1 + \frac{2\varepsilon\xi(2-\xi)}{(1-\xi)^2} \right] \tag{4.3}$$

where $S = \dfrac{R_0}{R} = 1-\xi, \xi = \dfrac{\delta}{R}, \varepsilon = \dfrac{\eta_1}{\eta_2} > 1$ and Q_0 is the Poiseuille flow rate for a liquid with viscosity η_1. The flow described by Equation (4.3) can be applied to a capillary in a porous media with relative permeability.

An analysis of Equation (4.3) demonstrates that the dependence of the liquid flow rate on the thickness of the near-wall layer $\delta = R - R_0$ is not monotonic and the maximum is reached at

$$\xi = 1 - \sqrt{\frac{\varepsilon}{2\varepsilon - 1}}$$

The complex flow rate is determined by the product

$$(1-\xi)^4 \left[\frac{2\varepsilon\xi(2-\xi)}{(1-\xi)^2} \right]$$

The former factor characterizes the cross-sectional area of the liquid flow while the latter determines the contribution of the near-wall layer with reduced viscosity and rises in the liquid flow rate. At a relatively small thickness of the near-wall layer the flow dominates and the liquid flow rate increases. At a relatively large thickness of the near-wall layer, the cross-sectional factor plays a major role and the flow rate decreases.

Let us assume that the near-wall layer is composed of gas alone. Then $\varepsilon = 65$ (the water-to-methane viscosity ratio at 308 K). As can be seen from Figure 4.1, although the dependence of the relative liquid flow rate on the dimensionless thickness of the near-wall gas layer qualitatively describes the experimental data, the maximum relative flow rate is higher than the experimental value by more than an order of magnitude.

This case is related to the fact that the near-wall layer possesses different properties from the bulk viscosity. Gas nuclei are formed at the surface of the pore channel and then migrate into the bulk liquid. It is reasonable to assume that at the contact line between the sphere

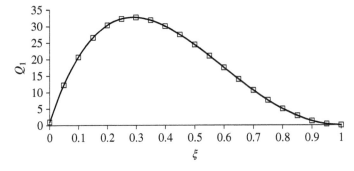

Figure 4.1 Theoretical dependence of the relative liquid flow rate on the dimensionless thickness of the near-wall gas layer.

segment-shaped gas nuclei (which cover the surface of the pore channel) and the liquid, the viscosity of the near-wall fluid is almost equal to the bulk liquid viscosity, while at the capillary wall, it is nearly equal to the viscosity of the gas. As a consequence, variations in the viscosity of the near-wall layer cab be described using the following exponential law:

$$\eta_b = \eta_2 \exp\left(\ln \varepsilon \frac{R-r}{R-R_0}\right) \tag{4.4}$$

where η_2 is the viscosity of the gas.

The steady-state flow of a viscous fluid in a capillary is known to be described by the following expression:

$$\frac{1}{r}\frac{d}{dr}\left(r\eta_b \frac{dv}{dr}\right) = -\frac{\Delta P}{l} \tag{4.5}$$

where η_b is determined by Equation (4.4).

The solution of Equation (4.5) yields the following expression for the fluid flow velocity in the near-wall layer:

$$v = -\frac{\Delta P}{2l\eta_2(\ln \varepsilon)^2}\left[(R-R_0)(R_0 - R + r\ln \varepsilon) \times \exp\left(\ln \varepsilon \frac{r-R}{R-R_0}\right)\right] \\ + \frac{a_{12}}{\eta_2} \exp\left(-\ln \varepsilon \frac{R}{R-R_0}\right) Ei\left(\ln \varepsilon \frac{r}{R-R_0}\right) + a_{22} \tag{4.6}$$

Under the above boundary conditions, based on Equations (4.1) and (4.5), the following formula for the liquid flow velocity can be derived:

$$v_1 = \frac{\Delta P R^2}{4l\eta_1}\left\{(1-\xi)^2 - \left(\frac{r}{R}\right)^2 + \frac{2\xi}{(\ln \varepsilon)^2}[\ln \varepsilon^\varepsilon \exp(1-\varepsilon) - (1-\xi)\ln \varepsilon \exp(1-\varepsilon)]\right\} \tag{4.7}$$

For the relative liquid flow, the following expression is produced:

$$Q_1 = \frac{Q}{Q_0} = (1-\xi)^4 \left\{1 + \frac{4\xi}{(1-\xi)^2(\ln \varepsilon)^2}[\ln \varepsilon^\varepsilon \exp(1-\varepsilon) - (1-\xi)\ln \varepsilon \exp(1-\varepsilon)]\right\} \tag{4.8}$$

An analysis of the produced dependencies shows that the nonmonotonic flow rate is determined by the product in the right-hand part of Equation (4.8). The inhomogeneity in the flow is produced by the fluid's low viscosity in the near-wall layer. At relatively small thicknesses of the near-wall layer, the latter factor prevails and the liquid flow rate grows. At relatively large thicknesses of the near-wall layer, the former factor prevails and the flow rate diminishes. The dependence of the relative liquid flow rate on the dimensionless thickness of the near-wall layer found at $\varepsilon = 65$ (Figure 4.2) is in good agreement with the experimental data, and, on the order of magnitude, the maximum liquid flow rate coincides with the experimental value.

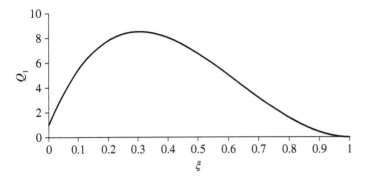

Figure 4.2 Theoretical dependence of the relative liquid flow rate on the dimensionless thickness of the near-wall layer.

4.1.1.2 Slip Effect

It was presumed that the capillary flow of a liquid in the presence of a motionless gas layer defined by Equation (4.1) is constrained by the following boundary conditions:

$$v = -b\frac{dv}{dr} \quad \text{at} \quad r = R_0$$
$$\frac{dv}{dr} = 0 \quad \text{at} \quad r = 0 \tag{4.9}$$

and the following relations for the flow velocity and flow rate were obtained:

$$v = \frac{\Delta P}{4l\eta}(R_0^2 - r^2) + b\frac{\Delta P R_0}{2l\eta}$$

$$Q = \frac{\pi \Delta P R_0^4}{8l\eta}\left(1 + \frac{4b}{R_0}\right)$$

The relative liquid flow rate is defined by the following equations:

$$Q_1 = \frac{Q}{Q_0} = \frac{R_0^4}{R^4}\left(1 + \frac{4b}{R_0}\right)$$

or

$$Q_1 = (1-\xi)^4\left[1 + \frac{4b}{R(1-\xi)}\right]. \tag{4.10}$$

This is now in line with references [15] and [16], where the liquid motion in the presence of a near-wall gas layer was considered according to Stokes, i.e. between two infinite parallel plates. The slip coefficient is expressed via the thickness of the near-wall layer and the capillary radius. Applying the Navier hypothesis of the velocity proportionality at a liquid–gas interface to the shear rate (it was actually used in Equation (4.9)), the following expression for the slip coefficient was obtained (the annular flow of a liquid and a gas is considered at a constant viscosity of the latter in Eq. (4.2)):

$$b = R\frac{\varepsilon\xi(2-\xi)}{2(1-\xi)} \tag{4.11}$$

If we substitute Equation (4.11) into Equation (4.10), we essentially get Equation (4.3). This demonstrates that an annular liquid or gas flow is identical to the slip flow (with a motionless boundary layer of the gas) when the slip coefficient is determined by Equation (4.11).

Then considering the case of a variable viscosity with allowance for the Navier hypothesis, the following relation derived for the slip coefficient can be produced:

$$b = R \frac{\xi}{(1-\xi)(\ln \varepsilon)^2}[\ln \varepsilon^\varepsilon \exp(1-\varepsilon) - (1-\xi)\ln \varepsilon \exp(1-\varepsilon)] \tag{4.12}$$

Then substituting Equation (4.12) into Equation (4.10), Equation (4.8) is achieved. Thus, the annular flow of a liquid- and a gas-saturated near-wall layer with a variable viscosity is identical to the slip flow at a motionless boundary layer, provided that the slip coefficient is determined by Equation (4.11). The maximum liquid flow rate takes place under the following condition:

$$\xi = 1 - \frac{3(a_0 + b_0) - \sqrt{9(a_0 + b_0)^2 - 8a_0(c_0 + 4b_0)}}{2(c_0 + 4b_0)}, \tag{4.13}$$

where $a_0 = \ln \varepsilon^\varepsilon \exp(1-\varepsilon)$, $b_0 = \ln \varepsilon \exp(1-\varepsilon)$, and $c_0 = (\ln \varepsilon)^2$. The estimations implemented through this formula at $\varepsilon=65$ suggest that the maximum flow rate is reached at $\xi \approx 0.3$.

The latter is clearly visible at theoretical dependences of the relative flow rate on the thickness of the near-wall layer at different capillary radii, as demonstrated in Figure 4.3 (the slip coefficient is determined by Equation (4.12)).

The flow dependencies show that, as the capillary radius decreases, the flow rate maximum shifts toward the wall. The maximum value obtained for the relative flow rate (≈ 8.5) should be considered to be an ultimately possible value that can be achieved in terms of hydrodynamics while the capillary surface is completely nullified by the near-wall gas-saturated layer. It is interesting to note that, in contrast to the experimental data, a reduction

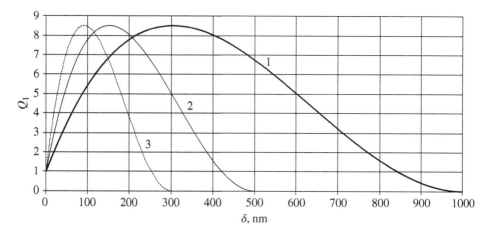

Figure 4.3 Theoretical dependences of the relative liquid flow rate on the near-wall layer thickness at different capillary radii: (1) 1.0, (2) 0.5, and (3) 0.3 μm.

in the capillary radius (or the permeability of a porous medium) does not cause growth at the maximum relative flow rate.

In a number of studies, it was found that even absolutely smooth hydrophobic surfaces are only partly covered with gas nanobubbles. Moreover, on a rough surface (with a nano-sized roughness), the number of formed nanobubbles is smaller and, as follows from reference [17], the degree of the lyophobic surface coverage with nanobubbles may be as low as 20%. It is obvious that, even with a lower surface coverage, bubbles are possible in a porous medium case.

The slip coefficient and liquid flow rate under conditions of an incomplete coverage of a capillary surface with gas phase bubbles (partial slip) can be found by applying the approach proposed earlier in reference [18]. The following equation is used to describe the liquid flow velocity:

$$v_c = fv_s + (1-f)v_0, \tag{4.14}$$

where v_s is the velocity of the liquid slip flow, v_0 is the Poiseuille liquid flow velocity, and f is the fraction of the capillary surface covered with gas bubbles. Applying the Navier hypothesis of the total velocity proportionality at a liquid–gas interface to the shear rate, the following relation is obtained from Equation (4.2) with regard to Equation (4.14) for the slip coefficient (in the case of the annular flow of a liquid and a gas at a constant viscosity of the latter):

$$b = R \frac{\varepsilon \xi (2-\xi)}{2(1-\xi)} [1 - f(1-\varepsilon)] \tag{4.15}$$

At $f = 1$, Equation (4.15) is transformed into Equation (4.11).

Considering the case of a variable viscosity of the near-wall layer with allowance for the Navier hypothesis and Equation (4.14), the following expression for the slip coefficient is derived from Equation (4.7):

$$b = R \left\{ \frac{(1-f)\xi(2-\xi)}{2(1-\xi)} + \frac{f\xi}{(1-\xi)(\ln \varepsilon)^2} [\ln \varepsilon^{\varepsilon} \exp(1-\varepsilon) - (1-\xi) \ln \varepsilon \exp(1-\varepsilon)] \right\} \tag{4.16}$$

At $f = 1$, Equations (4.16) and (4.12) are similar.

Equation (4.16) can be simplified by ignoring nonlinear ξ terms. In this case, the following equation is obtained:

$$b = R \left\{ \frac{\xi[(1-f)\ln \varepsilon + f(\varepsilon - 1)]}{(1-\xi)\ln \varepsilon} \right\}$$

Figure 4.4 shows the dependences of the near-wall layer thickness on the capillary radius at different slip coefficients. The slip coefficient on the near-wall layer thickness at different degrees of the pore channel surface coverage with gas phase nuclei $f(R = 10^{-6}$ m) is illustrated in Figure 4.5.

As can be seen from Figure 4.4, the thickness of the near-wall layer increases with the slip coefficient. As the capillary radius diminishes, the slip coefficient rises (at a fixed thickness of the gas-saturated near-wall layer). Moreover, at relatively small capillary radii (smaller than 200 nm), substantial variations in the thickness of the near-wall layer correspond to

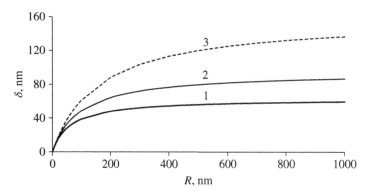

Figure 4.4 Theoretical dependences of the near-wall layer thickness on the capillary radius at different slip coefficients: (1) 1.0, (2) 1.5, and (3) 2 μm.

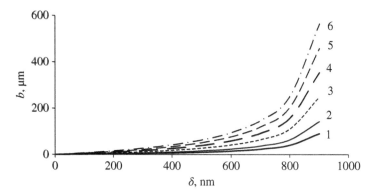

Figure 4.5 Theoretical dependences of the slip coefficient on the near-wall layer thickness at different degrees of pore channel surface coverage with the gas phase bubble f: (1) 0, (2) 0.2, (3) 0.4, (4) 0.6, (5) 0.8, and (6) 1.0.

small changes in the slip coefficient. It follows from Figure 4.5 that at a relatively small thickness of the near-wall layer, the slip coefficient does not vary significantly and begins to dramatically increase at $\delta > 800$ nm. At reduced thicknesses of the near-wall layer, higher f values are required to reach the same values of the slip coefficient.

For the case of the partial surface coverage with bubbles, the maximum flow rate takes place at

$$\xi = 1 - \frac{3f(a_0 + b_0) - \sqrt{9f^2(a_0 + b_0)^2 - 4(c_0 - fc_0 + 2fa_0)(2fc_0 - c_0 + 4fb_0)}}{2(2fc_0 - c_0 + 4fb_0)} \quad (4.17)$$

At $f = 1$, Equation (4.17) is transformed into equality (4.13).

The relative thickness of the near-wall layer corresponding to the maximum liquid flow rate is presented in Figure 4.6 as a function of the degree of the capillary surface coverage with gas bubbles. As can be seen from the figure, at $f > 0.15$, ξ does not vary significantly.

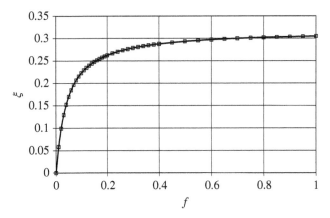

Figure 4.6 Theoretical dependence of the relative near-wall layer thickness corresponding to the maximum liquid flow rate on the degree of capillary surface coverage with gas bubbles.

When the radius of a capillary reduces, the specific surface area enlarges in accordance with the R_2/R_1 ratio, where R_2 and R_1 are the radii of the larger and smaller capillaries respectively. It is reasonable to assume that parameter f increases according to the same proportion as a decrease in the capillary radius, since the probability of the gas phase nucleation grows with the specific surface area. Assuming that $f = 0.1$ at $R = 1\,\mu\text{m}$, $f = 0.2$ at $R = 0.5\,\mu\text{m}$ and $f = 0.3$ at $R = 0.3\,\mu\text{m}$ were obtained. Figure 4.7 shows the dependences of the relative flow rate on the thickness of the near-wall layer at different capillary radii with allowance for the partial slip (here and below, $\varepsilon = 65$). According to the figure, a reduction in the capillary radius leads to an increase in the maximum relative flow rate, with the maximum shifting toward smaller thicknesses of the near-wall layer. The dependences of the relative flow rate on parameter ξ with regard to the partial slip are demonstrated in Figure 4.8 at different capillary radii. It follows from the figure that the maximum is observed at almost the same value of $\xi \approx 0.3$ irrespective of the capillary radius.

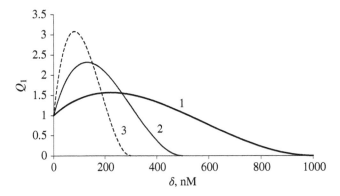

Figure 4.7 Theoretical dependences of the relative liquid flow rate on the near-wall layer thickness with an allowance for partial slip at different capillary radii: (1) 1.0, (2) 0.5, and (3) 0.3 μm.

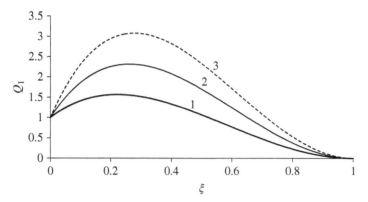

Figure 4.8 Theoretical dependences of the relative liquid flow rate on ξ with allowance for partial slip at different capillary radii: (1) 1.0, (2) 0.5, and (3) 0.3 μm.

The equation proposed in reference [19] had been used to find the dependence of the liquid flow rate on pressure. According to reference [9], $\beta \sim C$, where C is the volume concentration of gas phase nuclei. The dependence of the volume concentration of nuclei on pressure may then be written as follows [19]:

$$C = C_0 \exp\left(-\alpha \frac{P - P_c}{P_c}\right) \tag{4.18}$$

Here C_0 is the concentration of nuclei at $P = P_c$, where P_c is the saturation pressure and $\alpha = 1/f$ is a constant coefficient characterizing the degree of pore channel surface coverage with gas phase nuclei. The following expression was obtained from Equation (4.2):

$$\frac{P}{P_c} = 1 - f \ln \frac{C}{C_0}$$

Meanwhile, the volume fraction of the near-wall layer was determined as follows:

$$C = f \frac{\pi R^2 l - \pi R_0^2 l}{\pi R^2 l} = f\left[2\xi - \xi^2\right] \tag{4.19}$$

where l is the capillary length.

Consequently, using Equations (4.18) and (4.19), the dependence of the liquid flow rate on the pressure was derived from Equations (4.10) and (4.16) along with the relatively low pressure drop and constant pressure gradient along the liquid flow. Figure 4.9 illustrates the dependences of the relative flow rate on the pressure at different capillary radii. As can be seen from the figure, the proposed model satisfactorily describes the experimental data. Indeed, as the capillary radius (the permeability of a porous medium) decreases, the maximum of the liquid flow rate is reduced and shifts toward the saturation pressure. The pressure P_s corresponding to the onset of a rise in the liquid flow rate also decreases.

Let us estimate the volume concentration of the gas phase at the near-wall layer thickness corresponding to the maximum liquid flow rate ($\xi \approx 0.3$). It follows from Equation (4.19) that $C = f(2\xi - \xi^2) = 0.51f$; i.e. if the near-wall layer consists of gas alone and covers the entire

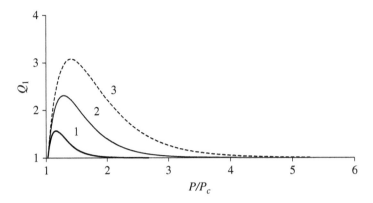

Figure 4.9 Theoretical dependences of the relative liquid flow rate on the pressure level at different capillary radii: (1) 1.0, (2) 0.5, and (3) 0.3 μm.

capillary surface $f = 0.05$–0.3 and the volume fraction of the gas phase will amount to 51%. Considering the fact that values correspond to the experimental data, the highest estimate of the volume concentration is $C = 1.6$–14%. As mentioned above, the near-wall layer represents a mixture of a liquid with densely packed gas nanobubbles; more-over, according to references [9], [10], and [11], the bubbles have the morphology of a sphere segment, with their curvature radii being noticeably larger than their heights. According to reference [10], the curvature radius was assessed as $R_c \approx 50\delta$; hence, as follows from Equation (3.9), the radius of the sphere segment base is $R_s \approx 10\delta$. The internal capillary surface, the surface area occupied by a single bubble, and the number of densely packed bubbles are determined respectively as

$$S_k = 2\pi R l, \quad S_b = \pi R_s^2 \quad \text{and} \quad N = \frac{S_k}{S_b} = \frac{2Rl}{R_s^2} \approx \frac{Rl}{50\delta^2}$$

According to Equation (3.10) the bubble volume is defined by $V_b = \frac{1}{3}\pi\delta^2(3R_c - \delta) \approx \frac{149}{3}\pi\delta^3$

The capillary volume is equal to $V_k = \pi R^2 l$. The volume concentration is found from the equation

$$C = f\frac{V_b N}{V_k} \approx f\frac{\delta}{R} = f\xi \tag{4.20}$$

Thus, when the entire capillary surface is covered with gas phase nuclei ($f = 1$) the volume concentration of the gas phase will amount to 30%. The lowest estimate for the volume concentration ($f = 0.05$–0.3) is $C = 0.9$–8%; i.e. a marked increase in the liquid flow rate is even possible at very low volume concentrations of the gas phase.

The effect of adding anionic surfactants for improving the wetting of the pore channel surface is determined via the following expression for the concentration of gas phase nuclei as a function of the work of heterogeneous nucleation:

$$C = \exp\left(-\frac{W_h}{KT}\right) \tag{4.21}$$

where W_h is the work of the heterogeneous nucleation of the gas phase. It follows from Equation (4.21) that the concentration of nuclei diminishes with a growth of heterogeneous nucleation of the gas phase (an increase in the surface wettability). The liquid flow rate decreases with a rise in the surface wettability, with other conditions being equal. It is also confirmed by the data reported in reference [20], where it was established that surfactant adsorption on a channel surface greatly (more than threefold) reduces the slip coefficient. According to the proposed model, the influence of surfactant additives can be considered by parameter f reduction in Equation (4.20).

4.1.1.3 Concluding Remarks

To summarize all of the above, it may be stated that the slip effect explains the experimental data available on the flow of gas-saturated liquids in a porous media. The mechanism of the observed phenomena is as follows. During the flow of a gasified (Newtonian or non-Newtonian) liquid characterized by the contact angle $\theta > 0°$, stable bubbles of a gas phase are formed on the surface of pore channels in the subcritical region. The bubble sizes are in a range of 1–100 nm and form a near-wall gaseous layer, which causes the liquid to slip on the surface and enhances its flow rate while decreasing the pressure to the saturation pressure. At the same time, while reducing the pressure, the fraction of the capillary cross-sectional area through which the liquid is transferred decreases because of the rise in bubble size. The competition between these two effects (see, for example, the analysis of Equation (4.8)) results in a nonmonotonic dependence of the liquid flow rate reduced to the Darcy flow rate (the relative flow rate) on the pressure. The relative flow rate reaches a maximum at $P/P_c = 1.1–1.5$. It is obvious that, before the maximum is reached, the slip effect predominates, while a subsequent decrease in the relative flow rate to 1 at $P/P_c = 1\,1$ is caused by the prevailing effect of a reduction in the effective cross-sectional area of pore channels due to an increase in the thickness of the gas-saturated near-wall layer. Note that the maximum of the relative flow rate is reached at a relatively low saturation of the pore volume with the gas, $C = 0.9–8\%$. A decline in the relative liquid flow rate at pressures below the saturation pressure, when it becomes lower than one, can be explained by the release of a free gas phase and the realization of the Jamin effect, which is caused by choking porous channels with macrosized gas phase bubbles.

This effect can take place at a relatively low saturation rate of the pore volume with gas. For example, according to reference [21], where the data are presented on liquid flow in microcapillaries containing gas bubbles that completely overlap the capillary cross-sectional area, the gas volume concentration is, in this situation, as low as 1.5%, while the liquid flow rate is twofold lower than the Poiseuille flow rate.

A rise in the contribution from the slip effect at a reduced permeability of a porous medium can be explained by an increase in the degree of pore channel coverage with gas phase bubbles, which grows with a rise in the specific surface area of a porous medium (see the text devoted to the data analysis and Figure 4.7). In this case, the maximum relative flow rate is reached at higher pressures, while the value of the relative flow rate at the point of the maximum increases. The enhancement of the slip coefficient with the liquid gas saturation is explained by a growth in the degree of pore channel coverage with gas phase bubbles.

A decrease in the slip coefficient upon the addition of an anionic surfactant, which increases the wettability of the pore channel surface, is explained by a reduction in the degree of pore channel surface coverage with gas phase bubbles. In this case the maximum relative flow rate is reached at lower pressures and the maximum relative flow rate reduces. Note that the assumption that gas phase bubbles adsorbed in a porous medium are motionless, while the liquid slips along the capillary surface. This assumption is supported by experimental data, which indicate that the flow of a gas–condensate mixture at pressures above the onset of condensation is accompanied by a noticeable rise in the gas flow rate (by approximately 50%). It follows from Equation (4.10) that if a gas condensate mixture flows in the annular regime, when a near-wall liquid phase is mobile, the gas flow rate must decrease. Hence, the increase in the gas flow rate observed in reference [22] can also be explained by the slip effect.

4.1.2 Flow of Gasified Newtonian Liquids in Porous Media at Reservoir Conditions

This section considers the isothermal flow of a gasified non-Newtonian liquid in a capillary and in a homogeneous/nonhomogeneous porous media at above saturation pressure conditions.

4.1.2.1 Fundamental Equations

When considering the flow of gassed liquids it is generally assumed that the Darcy equation for the liquid and gas can be expressed as follows:

$$mS_i v_i = -\frac{K_0 K_i}{\mu_i} \nabla P \tag{4.22}$$

where K_0, K_i, S_i, m, and μ_i are the absolute permeability of a porous medium, the relative permeability, the degree of saturation, porosity, and dynamic viscosity of the fluid respectively.

Equation (4.22) describes the flow in the porous medium when a liquid or gas moves in a system of continuous channels filled with any one of these fluids. However, such a concept is not true if the content of one of these phases (for example, the gas) is low [23] and the porous medium is mainly occupied by the liquid with isolate gas bubbles (nuclei).

It is known that bubble nucleation occurs mainly on the solid surface and is more intense at poorer surface wettability [17]. Some experimental works demonstrated that the adhesion of the liquid to the capillary surface is weakened in microcapillaries because of the gas adsorption. According to references [24] and [25], the gaseous interlayers and surface microbubbles may favor the liquid slipping along the capillary walls. Churaev, Sobolev, and Somov [26] note that the slip effect is related to the formation of gaseous bubbles on the capillary surface.

Hence, the gasified liquid at above saturation pressures can be considered to be a homogeneous incompressible fluid with apparent permeability that is determined by the slip. Then the solution to the stationary problem for a planar one-dimensional system is found by solving the following set of equations:

$$\frac{dV}{dx} = 0$$
$$V = -\frac{K(P)}{\mu}\frac{dP}{dx} \tag{4.23}$$

where $K(P)$ is the apparent permeability.

4.1.2.2 Apparent Permeability

For the flow with slip, the apparent permeability can by determined from expression [27]

$$K(P) = K_0\left(1 + \frac{4b}{R}\right) \tag{4.24}$$

where R is the average radius of the pore channel, which can be found from

$$R = \sqrt{\frac{8K_0}{m}}$$

and b is the slip coefficient. It is evident that the slip coefficient in this system is a function of pressure. For a relatively small pressure drop along the capillary length, Equation (4.24) can be expressed as

$$K(P) = K_0\left[1 + \frac{4b(P)}{R}\right]$$

The $b(P)$ function for the water–natural gas system was determined experimentally. The experiment was conducted by the following procedure:

- A water–gas mixture was prepared in a cell with a floating system; the gas concentration (in molar fractions) was 0.035 and the saturation pressure P_c was 3.0 MPa.
- A porous medium (with a length and diameter of 1 and 0.035 m respectively) composed of quartz sand (with a permeability of 0.15 µm^2 and a porosity of 0.2) was hydrophobized with a hydrocarbon liquid (the contact angle was 70°);
- The porous medium was evacuated, maintained at a constant temperature (303 K), and saturated with the gassed liquid at a pressure higher than the saturation pressure.
- The water–gas mixture was filtered at a constant pressure drop until a constant flow rate was attained at different pressure levels ($P_0 = 2$–$3.5\,P_c$ and $P_e = 1$–$2.5\,P_c$, where P_0 and P_e are the pressures at the inlet and outlet of the porous medium respectively).

The experiment was performed at a relatively small pressure drop equal to 0.25 MPa (less than 10% of the total pressure).

The slip coefficient at the constant pressure was determined by the expression

$$b|_P = \left(\frac{Q}{Q_0} - 1\right)\frac{R}{4}$$

where Q and Q_0 are the liquid flow rates with and without slip respectively and $P = (P_0 + P_e)/2$. Figure 4.10 demonstrates the nonmonotonic pressure drop of the slip coefficient.

The observed dependence can be explained by the following kinetic mechanism. The slip was observed between $P_S \approx 2.6$ MPa and $P_C = 7.8$ MPa. When the pressure is lowered from

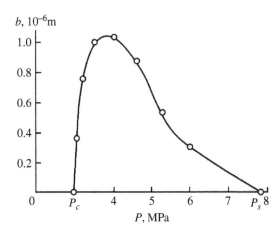

Figure 4.10 Experimental pressure dependence of the slip coefficient.

the high end, the volume concentration of gas bubbles at the solid surface of pore channels rises and the slipping coefficient increases continuously. The maximum is observed at pressure $P \approx 1.3 P_c$. However, the bubble radius also increases [3] and the slip coefficient begins to decrease when this radius becomes comparable with the mean radius of capillaries in the actual porous medium.

The experimental dependence is described by the following empirical equation:

$$b(P) = b_0 \left\{ a \left(\frac{P - P_c}{P_s - P_c} \right) \exp \left[-c \left(\frac{P - P_c}{P_s - P_c} \right) \right] \right\}$$

where $b_0 = 1.1 \times 10^{-6}$ m is the slip coefficient at $P \approx 1.3 P_c$ (i.e. the maximal value), P_s is the pressure corresponding to the onset of the slip ($P_s \approx 2.6 \, P_c = 7.8$ MPa), and $a \approx 16$ and $c \approx 5.8$ are numerical constants. Therefore, the apparent permeability for flow with slip can be expressed as

$$K(P) = K_0 \left\{ 1 + \frac{4 b_0}{R} \left[a \frac{P - P_c}{P_s - P_c} \exp \left(-c \frac{P - P_c}{P_s - P_c} \right) \right] \right\} \qquad (4.25)$$

4.1.2.3 Steady-State Flow

Firstly, let us consider the steady-state flow at $P_0 > P_s > P_e > P_c$ (shown in Figure 4.11a). Solving the set of equations (4.23), it can be found that

$$\frac{K(P)}{\mu} \frac{dP}{dx} = V_S \qquad (4.26)$$

where V_S is the flow rate at point $x = x_S$ that corresponds to the pressure at which the liquid slip begins.

The pressure distribution can be found from Equation (4.26), where

$$\int_{P_s}^{P} K(P) \, dP = \mu V_S \int_0^x dx \qquad (4.27)$$

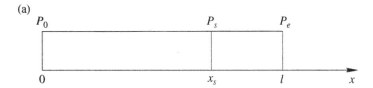

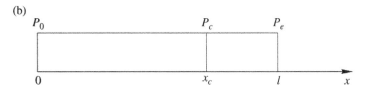

Figure 4.11 Diagram of the flow flux in the porous medium at (a) $P_0 > P_s > P_e > P_c$ and (b) $P_s \geq P_0 > P_c > P_e$.

At $x = l$ (l is the length of the porous sample), the pressure is equal to P_0. At $x_c < x < l$, pressure P is larger than P_S and flow without slip takes place, i.e. $K(P) = K_0$. It then follows from Equation (4.27) that

$$\int_{P_S}^{P_0} K_0 P = \mu V_S \int_{x_s}^{l} dx \tag{4.28}$$

At $x_e < x < x_s$, where x_s corresponds to pressure P_S, flow with slip occurs, and it follows from Equation (4.27) that

$$\int_{P_e}^{P_S} K(P)\, dP = \mu V_S \int_{x0}^{x_s} dx \tag{4.29}$$

Then, from Equations (4.28) and (4.29), the following expression can be derived:

$$x_S = l \left[1 + \frac{K_0(P_0 - P_S)}{P_S \int_{P_e}^{P_S} K(P)\, dP} \right]^{-1}$$

According to the slip from Equation (4.28), the liquid flow rate is

$$Q = \frac{K_0(P_0 - P_S)}{\mu(l - x_S)} F$$

where F is the cross-sectional area of the porous medium.

According to the Darcy law, the liquid flow rate is equal to

$$Q_0 = \frac{K_0(P_0 - P_e)}{\mu l} F$$

Thus, the following expression for the dimensionless liquid flow rate was obtained, where $Q_1 = Q/Q_0$:

$$Q_1 = \frac{(P_0 - P_S)l}{(P_0 - P_e)(l - x_S)}. \tag{4.30}$$

At $P_S > P_0 > P_e > P_c$, it follows from Equation (4.27) that the liquid flow rate accounting for the slip is

$$Q = \frac{\int_{P_e}^{P_0} K(P)\, dP}{\mu l} F$$

and the dimensionless flow rate is expressed as

$$Q_1 = \frac{\int_{P_e}^{P_0} K(P)\, dP}{k_0(P_0 - P_e)} \tag{4.31}$$

Opening the integral $\int_{P_1}^{P_2} K(P)\, dP$ and accounting for Equation (4.25), the following equation is obtained:

$$\int_{P_1}^{P_2} K(P)\, dP = K_0 \left\{ (P_2 - P_1) + \frac{4b_0}{R} \frac{a}{c^2} (P_S - P_c) \right.$$
$$\times \left[\exp\left(-c\frac{P_1 - P_c}{P_S - P_c}\right)\left(1 + c\frac{P_1 - P_c}{P_S - P_c}\right) \right.$$
$$\left.\left. - \exp\left(c\frac{P_2 - P_c}{P_S - P_c}\right)\left(1 + c\frac{P_1 - P_c}{P_S - P_c}\right)\right]\right\}$$

The available experimental data [1, 19] agree well with the dependences obtained.

The points in Figure 4.12 represent the experimental data from reference [1] for the artificial oil–natural gas system. The solid curve represents the results of calculations from formulas (4.30) and (4.31) at $K_0 = 0.15 \times 10^{-12}$ m^2, $m = 0.2$, $b_0 = 0.4 \times 10^{-6}$, $a \approx 16$, and $c \approx 5.8$. As seen from Figure 4.12 the proposed model accurately describes the experimental data.

Figure 4.13 represents the calculated dependence (the solid line) of the dimensionless flow rate on the pressure drop ΔP for the n-hexane-carbon dioxide system in comparison with the experimental data [1]. The calculations were performed using Equation (4.30) at $K_0 = 0.35 \times 10^{-12}$ m^2, $m = 0.2$, $b_0 = 5.9 \times 10^{-6}$, $a \approx 16$, and $c \approx 5.8$. The figure demonstrates that the quantitative description is also consistent with the proposed model.

Figure 4.14 demonstrates the pressure distribution $P(x)$ calculated for the experimental data [28] at various pressures at the outlet of the porous medium. It is shown that when the slip due to intense nucleation at the walls of pore channels takes place then the slope

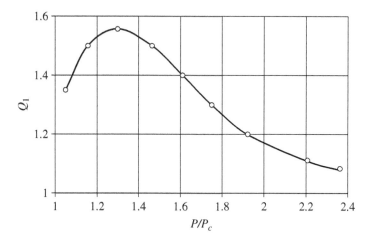

Figure 4.12 Dimensionless liquid flow rate as a function of the dimensionless pressure drop. The points present the experimental data [1]. The solid curve is calculated by formulas (4.30) and (4.31).

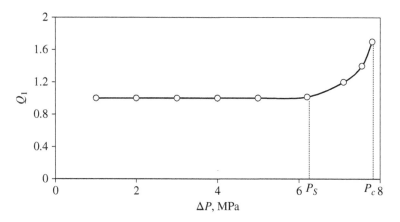

Figure 4.13 Dimensionless liquid flow rate as a function of the pressure drop at $P_0 > P_s > P_e > P_c$. The points present the experimental data [1]. The solid curve is calculated by formula (4.30).

of the pressure distribution curve decreases substantially in the pressure interval $P_s > P_e$. This process leads to an abnormal increase in the liquid flow rate.

Note that nucleation at the walls of pore channels and, correspondingly, the effect of the slip depends on the wettability of the porous medium. Suleimanov and Azizov [19] have reported experimental data on the influence of the wettability of the porous medium on the flow of the subcritical gassed liquid. These data demonstrated that the phenomenon of the abnormal increase in the liquid flow rate near the saturation pressure vanished when the porous medium wettability was improved. The wettability can be calculated by assuming in Equation (4.25) that $c = c_1 f(\theta)$, where c is a constant and $f(\theta)$ is a function of the contact angle θ [19].

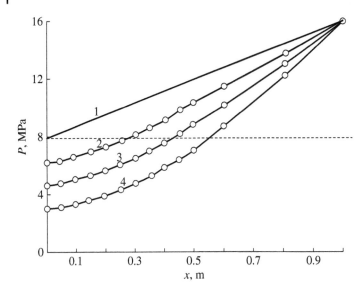

Figure 4.14 Pressure distribution P(x) at $P_0 > P_s > P_e > P_c$ and at the pressure at the outlet of the porous medium equal to (1) 7.8, (2) 6.2, (3) 4.6, and (4) 3.0 MPa.

It was assumed that a system has lower pressure at the outlet of the porous medium compared to the saturation pressure $P_S \geq P_0 > P_c > P_e$ (a diagram in Figure 4.11b). Using Equation (4.27), for $x_c < x < l$ the following expression is obtained:

$$\int_{P_e}^{P_0} K(P)\, dP = \mu V_S \int_{x_c}^{l} dx \tag{4.32}$$

According to reference [29], at $0 < x < x_c$,

$$K(P) = K_0 \left(\frac{P}{P_c}\right)^n$$

where $n = 0.42$. Then

$$\int_{P_e}^{P_c} K_0 \left(\frac{P}{P_c}\right)^n dP = \mu V_S \int_0^{x_c} dx \tag{4.33}$$

Combining Equations (4.32) and (4.33), it can be found that

$$x_c = l \left[1 + \frac{(n+1) P_c^n \int_{P_c}^{P_0} K(P)\, dP}{K_0 \left(P_c^{n+1} - P_e^{n+1}\right)} \right]^{-1}$$

The liquid flow rate is expressed as

$$Q = \frac{K_0(P_0 - P_c)}{\mu(l - x_c)} F$$

Correspondingly, the dimensionless liquid flow rate is

$$Q_1 = \frac{(P_0 - P_c)l}{(P_0 - P_c)(l - x_c)} \tag{4.34}$$

Figure 4.15 represents the dimensionless liquid flow rate as a function of the pressure drop (e.g. differential). Calculations were performed by formulas (4.31) and (4.34). The principal parameters were taken as equal to those reported in reference [28]. The calculated dependence agrees qualitatively with the experimental data [1]. It is shown that when the pressure at the outlet of the porous medium is lower than the saturation pressure, the dimensionless rate decreases dramatically and attains values less than unity, although the pressure at the inlet of the porous medium is in the range corresponding to the significant slip of the liquid. Figure 4.16 represents the pressure distribution P(x) along the length of the porous sample at different pressures at the outlet of the porous medium. The curves become steeper (e.g. the flow resistance is increased) when the pressure at the outlet of the porous medium becomes lower than the saturation pressure. The reason is a dramatically high increase in porous medium gas saturation that is accompanied by the Jamin effect.

Hence, the abnormal increase in the flow rate of the gasified liquid near (above) the saturation pressure can be explained by the slip effect of the liquid due to intense subcritical nucleation at the walls of pore channels. Note that it is especially interesting to study the flow of the gassed non-Newtonian liquids with slipping. This issue will be discussed in the next section.

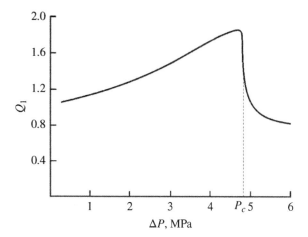

Figure 4.15 Dimensionless liquid flow rate as a function of the pressure drop at $P_s \geq P_0 > P_c > P_e$. Calculations were performed by formulas (4.31) and (4.34).

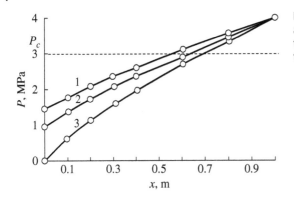

Figure 4.16 Pressure distribution P(x) at $P_S \geq P_0 > P_c > P_e$ and at the pressure at the outlet of the porous medium P_e equal to (1) 1.5, (2) 1.0, and (3) 0.0 MPa.

4.2 Hydrodynamics of Nanogas Emulsions in Heavy Oil Reservoirs

4.2.1 Flow Mechanism of Gasified Non-Newtonian Liquids

Generally, heavy oils demonstrate non-Newtonian rheological behavior. In this section, the mechanism of the slip has been considered for a gasified non-Newtonian liquid flowing in a capillary at a pressure higher than the saturation pressure (under conditions of subcritical gas phase nucleation). The most commonly applied power law has been used as the mathematical model for the non-Newtonian liquid flow. Models have been proposed for the flow of a power-type (Ostwald–de Waele) liquid in the presence of its slip on a capillary surface.

4.2.1.1 Annular Capillary Flow Scheme

Let us assume that gas bubbles are formed on a capillary surface and are mobile. As a result, a boundary layer of a liquid saturated with gas bubbles has a lower viscosity than the central part of the liquid. It is reasonable to assume that, as the pressure is reduced to the saturation pressure, the work of nucleation is reduced and the bubbles grow, which may result in thickening of the boundary layer with decreased viscosity.

Let us investigate the annular flow of a non-Newtonian (power-type) liquid in a capillary with a boundary gas layer. Allowing for the finiteness of stress τ, in the center of the capillary, at $r = 0$, the velocity of the liquid is determined by the following equation [30]:

$$v_1 = -\frac{n}{n+1}\left(\frac{\Delta P}{2\eta_0 l}\right)^{\frac{1}{n}} r^{\frac{1}{n}+1} + a_{21} \tag{4.35}$$

The flow velocity of the boundary gas layer is found from the relation

$$v_2 = -\frac{\Delta P}{4l\eta_2} r^2 + a_{12} \ln r + a_{22} \tag{4.36}$$

where η_0 is a constant ($\eta_0 = \tau/\gamma^n$). The constants of integration are determined from the following boundary conditions: $v_2 = 0$, $r = R$, and $\tau_2 = \tau_1$, $v_1 = v_2$, $r = R_0$.
The flow velocity is found from Equations (4.35) and (4.36) as follows:

$$v_1 = \frac{n}{n+1}\left(\frac{\Delta P}{2\eta_0 l}\right)^{\frac{1}{n}} R_0^{\frac{1}{n}+1}\left[1-\left(\frac{r}{R_0}\right)^{\frac{1}{n}+1}\right] + \frac{\Delta P R^2}{4\eta_2 l}\left[1-\left(\frac{R_0}{R}\right)^2\right] \qquad (4.37)$$

while the relative flow rate is determined from Equation (4.37) as

$$Q_1 = \frac{Q}{Q_0} = (1-\xi)^{\frac{1}{n}+3}\left[1 + \frac{3n+1}{2n}\frac{\varepsilon\xi(2-\xi)}{(1-\xi)^2}\right] \qquad (4.38)$$

$$\varepsilon = \frac{\eta_1}{\eta_2}, \eta_1 = \eta_0\left(\frac{\Delta P R_0}{2\eta_0 l}\right)^{1-\frac{1}{n}} \qquad (4.39)$$

where η_1 is the viscosity of the liquid at $r = R_0$, i.e. at the liquid–gas interface, and

$$Q_0 = \frac{\pi n}{3n+1}\left(\frac{\Delta P}{2\eta_0 l}\right)^{\frac{1}{n}} R^{\frac{1}{n}+3}$$ is the flow rate of a power-type liquid with viscosity

$$\eta_1 = \eta_0\left(\frac{dv_1}{dr}\right)^{n-1}$$

At $n = 1$, Equations (4.37) and (4.38) are transformed into the corresponding equations for Newtonian liquids (see reference [31] and Equations (4.36) and (4.37)). Analysis of Equation (4.38) shows that the dependence of the liquid flow rate on the boundary layer thickness $\delta = R - R_0$ is also nonmonotonic, with the maximum flow rate being reached at a dimensionless thickness of the boundary layer ($\xi = \delta/R$) equal to

$$\xi = 1 - \sqrt{\frac{(n+1))\varepsilon}{(3n+1)\varepsilon - 2n}} \qquad (4.40)$$

At $n = 1$, Equation (4.40) is transformed into the equality

$$\xi = 1 - \sqrt{\frac{\varepsilon}{2\varepsilon - 1}}$$

Analysis of Equation (4.40) shows that ξ increases with n, other conditions being constant. Given this, at $n \to 0, \xi \to 0$, while at $n \to \infty$, $\xi \to 0.42$. At the intermediate values $n = 0.5 - 2$, $\xi = 0.22 - 0.34$ at $\varepsilon \geq 30$.

For the above-mentioned reasons, the non-monotony of the flow rate is determined by the product

$$(1-\xi)^{\frac{1}{n}+3}\left[\frac{3n+1}{2n}\frac{\varepsilon\xi(2-\xi)}{(1-\xi)^2}\right]$$

Figure 4.17 illustrates the dependences of the relative liquid flow rate on the dimensionless thickness of the boundary gas layer at different n values. It is obvious that dependences qualitatively describe the obtained experimental data and the maximum value of the relative flow rate markedly exceeds it.

The annular flow of a power-type liquid was considered with a variable viscosity of the boundary layer in a capillary (as a first approximation, it was assumed that the liquid in the boundary layer is Newtonian). By solving the corresponding above-mentioned equation

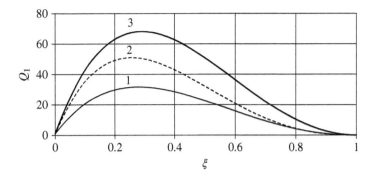

Figure 4.17 Relative liquid flow rate vs. dimensionless thickness of the boundary gas layer dependences at different n values: (1) 1.0, (2) 1.5, and (3) 0.9; $\Delta P/l = 1$ MPa/m.

under the same boundary conditions, the following relation was obtained for the liquid velocity:

$$v_1 = \frac{n}{n+1}\left(\frac{\Delta P}{2\eta_0 l}\right)^{\frac{1}{n}} R^{\frac{1}{n}+1}\left\{(1-\xi)^{\frac{1}{n}+1} - \left(\frac{r}{R}\right)^{\frac{1}{n}+1}\right.$$
$$\left. + \frac{n+1}{n}\frac{\xi}{(1-\xi)^{1-\frac{1}{n}}(\ln \varepsilon)^2}[\ln \varepsilon^\varepsilon \exp(1-\varepsilon) - (1-\xi)\ln \varepsilon \exp(1-\varepsilon)]\right\}$$

(4.41)

The following expression for the relative flow rate was obtained from Equation (4.41):

$$Q_1 = (1-\xi)^{\frac{1}{n}+3}\left\{1 + \frac{3n+1}{n}\frac{\xi}{(1-\xi)^2(\ln \varepsilon)^2}[\ln \varepsilon^\varepsilon \exp(1-\varepsilon) - (1-\xi)\ln \varepsilon \exp(1-\varepsilon)]\right\}$$

(4.42)

At $n = 1$, Equations (4.41) and (4.42) are transformed into corresponding expressions for Newtonian liquids (see reference [31] and Equations (4.41) and (4.42)).

For the above-described reasons, the nonmonotony of the flow rate is determined by the product

$$(1-\xi)^{\frac{1}{n}+3}\left\{\frac{3n+1}{n}\frac{\xi}{(1-\xi)^2(\ln \varepsilon)^2}[\ln \varepsilon^\varepsilon \exp(1-\varepsilon) - (1-\xi)\ln \varepsilon \exp(1-\varepsilon)]\right\}$$

4.2.1.2 Slip Effect

If we now assume that nucleation mainly occurs on the solid surface of capillaries in a porous medium and the formed (adsorbed) bubbles are retained by the porous medium and are, for the most part, motionless [32].

It was considered that the motion of a non-Newtonian (power-type) liquid in a capillary in the presence of a boundary gas layer is analyzed by using the determining constant a_{21} from Equation (4.35) under the conditions $v = -b(dv/dr)$, $r = R_0$. It was found that the following relations for the flow velocity and flow rate respectively are valid:

$$v = \left(\frac{n}{n+1}\right)\left(\frac{\Delta P}{2\eta_0 l}\right)^{\frac{1}{n}} R_0^{\frac{1}{n}+1}\left[1-\left(\frac{r}{R_0}\right)^{\frac{1}{n}+1} + \frac{n+1}{n}\frac{b}{R_0}\right]$$

$$Q = \left(\frac{\pi n}{3n+1}\right)\left(\frac{\Delta P}{2\eta_0 l}\right)^{\frac{1}{n}} R_0^{\frac{1}{n}+3}\left[1+\frac{3n+1}{n}\frac{b}{R_0}\right]$$

The expression for the relative liquid flow rate acquires the following form:

$$Q_1 = \frac{Q}{Q_0} = \left(\frac{\pi n}{3n+1}\right)\left(\frac{\Delta P}{2\eta_0 l}\right)^{\frac{1}{n}}\left(\frac{R_0}{R}\right)^{\frac{1}{n}+3}\left[1+\frac{3n+1}{n}\frac{b}{R_0}\right]$$

or

$$Q_1 = (1-\xi)^{\frac{1}{n}+3}\left[1+\frac{3n+1}{n}\frac{b}{R(1-\xi)}\right] \tag{4.43}$$

By applying the Navier hypothesis to Equation (4.37), the slip coefficient was expressed via the thickness of the boundary layer and the capillary radius can be obtained by the equation as presented in reference [31] (Equation (4.13)), where ε is defined by relation (4.39).

In the case of variable viscosity, by applying the Navier hypothesis to Equation (4.41), the equation reported in reference [31] for the slip coefficient (Equation (4.14)) was obtained, where ε is also determined by Equation (4.39). In both cases, the solution coincides with that for Newtonian liquids. Substituting expressions for slip coefficients into Equation (4.43), Equations (4.38) and (4.42) can be derived. Thus, for the case of a non-Newtonian liquid, the annular flows of the liquid and the boundary layer are identical to the slip flow with a motionless boundary layer, when the slip coefficient is determined by equations presented in reference [31] (Equations (4.13) and (4.14)).

The maximum liquid flow rate is reached under the following condition:

$$\xi = 1 - \frac{\frac{2n+1}{n}(a_0+b_0) - \sqrt{\left(\frac{2n+1}{n}\right)^2(a_0+b_0)^2 - 4\frac{n+1}{n}a_0\left(c_0+\frac{3n+1}{n}b_0\right)}}{2\left(c_0+\frac{3n+1}{n}b_0\right)}$$

(4.44)

where $a_0 = \ln \varepsilon^\varepsilon \exp(1-\varepsilon)$, $b_0 = \ln \varepsilon \exp(1-\varepsilon)$, $c_0 = (\ln \varepsilon)^2$, and ε is determined by Equation (4.39). At $n=1$, Equation (4.44) is transformed into a corresponding equation for Newtonian liquids (see reference [31] and Equation (4.15)). Analysis of expression (4.10) shows that ξ increases with n, other conditions being equal. Given this, at $n \to 0$, $\xi \to 0$, while at $n \to \infty$, $\xi \to 0.5 (\varepsilon \to \infty)$. For values $n = 0.5 \div 2$, $\xi = 0.23 \div 0.37$ at $\varepsilon \approx 30 \div 200$.

In the case under consideration, the viscosity of the liquid at the liquid–boundary layer interface depends on the pressure gradient (see Equation (4.39)). Therefore, the relative liquid flow rate also depends on the pressure gradient, with other conditions being equal. The dependences of the maximum relative liquid flow rate and the boundary layer thickness correspond to the relative flow rate maximum on the pressure gradient. They are illustrated in Figures 4.18 and 4.19 respectively for different n values.

As can be seen from the figures, the amplitude of the relative liquid flow rate maximum and the boundary layer thickness that corresponds to this maximum decrease for pseudoplastic and increase for dilatant liquids with a rise in the pressure gradient. This is based on

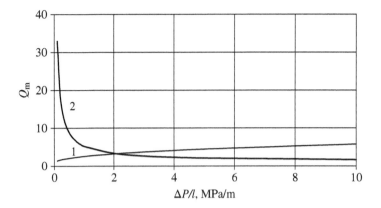

Figure 4.18 Relative liquid flow rate vs. pressure gradient dependences at different n values: (1) 2.0 and (2) 0.5.

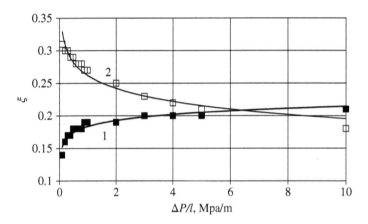

Figure 4.19 Relative thickness of the boundary layer corresponding to the maximum liquid flow rate vs. pressure gradient dependences at different n values: (1) 2.0 and (2) 0.5.

the fact that, for pseudoplastic liquids, the viscosity at the liquid–boundary layer interface decreases with an increase in the pressure gradient, while for dilatant liquids it increases; i.e. as the pressure gradient rises, in the first case, the average viscosity of the liquid tends to the average viscosity of the boundary layer, while, in the second case, it deviates from the latter.

The coefficient of the slip under conditions of its partial realization is found from the equation $v_c = f v_s + (1-f) v_0$, where v_s is determined by Equation (4.37), while v_0 is, in this case, the flow velocity of a power-type liquid without slip:

$$v_0 = \left(\frac{n}{n+1}\right)\left(\frac{\Delta P}{2\eta_0 l}\right)^{\frac{1}{n}} R^{\frac{1}{n}+1}\left[1-\left(\frac{r}{R}\right)^{\frac{1}{n}+1}\right] \tag{4.45}$$

Using the Navier hypothesis, the following relation for the case of a constant viscosity of the boundary layer was obtained:

$$b = R \left\{ f \frac{\varepsilon \xi (2-\xi)}{2(1-\xi)} + (1-f) \left(\frac{n}{n+1} \right) \left[\frac{1 - (1-\xi)^{\frac{1}{n}+1}}{(1-\xi)^{\frac{1}{n}}} \right] \right\} \quad (4.46)$$

where ε is determined by Equation (4.39). At $n = 1$, Equation (4.46) is transformed into the corresponding equation for Newtonian liquids (see reference [31] and Equation (4.17)).

For the case of variable viscosity of the boundary layer,

$$b = R \left\{ \left(\frac{n}{n+1} \right) \frac{(1-f)\left[1 - (1-\xi)^{\frac{1}{n}+1}\right]}{(1-\xi)^{\frac{1}{n}}} \right.$$

$$\left. + \frac{f\xi}{(1-\xi)(\ln \varepsilon)^2} [\ln \varepsilon^\varepsilon \exp(1-\varepsilon) - (1-\xi)\ln \varepsilon \exp(1-\varepsilon)] \right\} \quad (4.47)$$

where ε is determined by Equation (4.39). Equation (4.47) is transformed into the corresponding equation for Newtonian liquids (see reference [31] and Equation (4.18)).

Analysis of Equation (4.47) shows that the pattern of the $\delta(R)$ and $b(\delta)$ dependences for a power-type liquid remains unchanged and corresponds to similar dependences for Newtonian liquids (see reference [31] and Figures 4.20 and 4.21). Since ε of a power-type liquid depends on the pressure gradient, the slip coefficient of pseudoplastic liquids at low-pressure gradients (lower than 2 MPa/m) is higher compared to dilatant liquids. Meanwhile, at high-pressure gradients the slip coefficient is lower compared to dilatant liquids. Note for the both mentioned cases all other conditions should be equal.

With allowance for a partial slip, the value of ξ corresponding to the peak of the flow rate can be found at different parameters from the following expression:

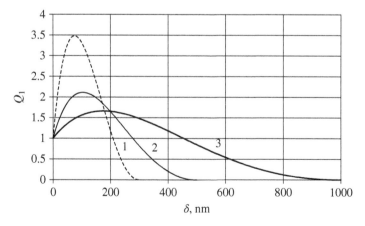

Figure 4.20 Relative flow rate Q_1 vs. boundary layer thickness dependences at different capillary radii and partial slip for a pseudoplastic liquid: $1-R = 0.3$, $2-0.5$, $3 = (1)\ 0.3$, $(2)\ 0.5$, and $(3)\ 1.0\ \mu m$; $n = 0.5$; and $\Delta P/l = 10$ MPa/m.

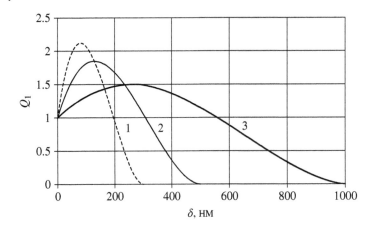

Figure 4.21 Relative flow rate Q_1 vs. boundary layer thickness dependences at different capillary radii and partial slip for dilatant liquid: R = (1) 0.3, (2) 0.5, and (3) 1.0 µm; n = 2; and $\Delta P/l$ = 1 MPa/m.

$$\left[\frac{c_0}{n} + \frac{3n+1}{n^2}b_0 f - \frac{3n+1}{n+1}c_0(1-f)\right](1-\xi)^2 - \frac{2n+1}{n^2}(a_0+b_0)f(1-\xi)$$

$$+ \frac{2}{n+1}c_0(1-f)(1-\xi)^{1-\frac{1}{n}} + \frac{n+1}{n^2}a_0 f = 0 \qquad (4.48)$$

where $a_0 = \ln \varepsilon^\varepsilon \exp(1-\varepsilon)$, $b_0 = \ln \varepsilon \exp(1-\varepsilon)$, $c_0 = (\ln \varepsilon)^2$, and ε is determined by Equation (4.39). At $n = 1$, Equation (4.48) is transformed into the corresponding relation for Newtonian liquids (see reference [31] and Equation (4.19)).

Dependences of the relative flow rate on the boundary layer thickness at different capillary radii, with allowance for a partial slip (by analogy with a Newtonian liquid, f = 0.1 at R = 1 µm, f = 0.2 at R = 0.5 µm, and $f \approx 0.3$ at R = 0.3 µm), are shown in Figures 4.20 and 4.21 for n = 0.5 and 2 respectively. As can be seen from the figures, a similar behavior is observed for the power-type liquid as for the Newtonian one (see reference [31] and Figure 4.7).

4.2.2 Flow of Gasified Non-Newtonian Liquids in Porous Media at Reservoir Conditions

The following sections consider the isothermal flow of a gasified non-Newtonian liquid in a capillary and porous media (i.e. homogeneous/nonhomogeneous) at above saturation pressure conditions. The power law model was used to define the flow of a non-Newtonian liquid. The stationary solution for a power-law liquid flow with slip in a capillary and a porous medium was found.

4.2.2.1 Capillary Flow

Let us regard a power-law liquid flow with slip in a cylindrical capillary. It is known that the relation between the shear stress and shear rate for power-law liquids is described by the expression

$$\tau = k_0 \gamma^n$$

where k_0 is a constant. The flow rate of the power-law liquid in a capillary is described by the following equation [30]:

$$v = -\left(\frac{n}{n+1}\right)\left(\frac{\Delta P}{2k_0 l}\right)^{\frac{1}{n}} r^{\frac{1}{n}+1} + C \qquad (4.49)$$

where C is a constant. Using the boundary condition that takes the slip effect into account,

$$v_R = -b\left(\frac{dv}{dr}\right)_R$$

and solving Equation (4.49), we can find the constant value:

$$v = \left(\frac{n}{n+1}\right)\left(\frac{\Delta P}{2k_0 l}\right)^{\frac{1}{n}} R^{\frac{1}{n}+1}\left(1 + \frac{b}{R}\frac{n+1}{n}\right) - \left(\frac{n}{n+1}\right)\left(\frac{\Delta P}{2k_0 l}\right)^{\frac{1}{n}} r^{\frac{1}{n}+1}$$

The volume flow rate is determined by

$$Q = 2\pi \int_0^R vr\, dr = \frac{\pi n}{3n+1}\left(\frac{\Delta P}{2k_0 l}\right)^{\frac{1}{n}} R^{\frac{1}{n}+3}\left(1 + \frac{3n+1}{n}\frac{b}{R}\right)$$

with the correction for a slip effect:

$$\frac{Q}{Q_0} = \left(1 + \frac{3n+1}{n}\frac{b}{R}\right).$$

4.2.2.2 Flow in a Homogeneous Porous Medium

In the case of flow in a porous medium, according to the Cozeny–von Karman equation [33],

$$v = -B\left|\frac{dP}{dx}\right|^{\frac{1}{n}-1}\frac{dP}{dx} \qquad (4.50)$$

where

$$B = B_0\left(1 + \frac{3n+1}{n}\frac{b}{R}\right)$$

is the coefficient of flow with allowance for the slip effect and $B_0 = \frac{n}{3n+1}$ $k_0^{-\frac{1}{n}} 2^{\frac{1}{2}(3+\frac{1}{n})} m^{-\frac{1}{2}(1+\frac{1}{n})} k^{\frac{1}{2}(1+\frac{1}{n})}$ (where m is the porosity and k is the permeability of a porous medium) is the coefficient of the flow without slip that is determined by conventional passage from a capillary flow to flow in a porous medium.

Taking into account the pressure dependence of parameter B, the flow will be reduced to the pressure dependence of the slip coefficient

$$B(p) = B_0\left(1 + \frac{3n+1}{n}\frac{b(p)}{R}\right)$$

For dependence $b(p)$, it is possible to use experimental data [34]:

$$b(p) = b_0 \left(a \frac{P - P_c}{P_S - P_c} \exp\left[-c\left(\frac{P - P_c}{P_S - P_c}\right)\right]\right) \qquad (4.51)$$

(Note that the slip effect can also be taken into account using the apparent permeability from Equation (4.50) as outlined in reference [34]). Hereafter the designations accepted in reference [34] will be used.

We will now consider one-dimensional steady-state flow in a cylindrical porous specimen when $P_S > P_0 > P_e > P_c$ (here P_S is the pressure at which the slip begins, P_0 and P_e are the pressures at the inlet and outlet of the porous medium, and P_c is the pressure of saturation of the liquid with the gas). It was expected that the slip effect takes place throughout the porous medium (see Figure 4.22a).

Solving Equation (4.50) together with the equation of continuity for an incompressible liquid,

$$\frac{dv}{dx} = 0 \qquad (4.52)$$

and after some transformations, we obtain an expression for the rate of flow with slip:

$$v = \left(-\frac{1}{l} \int_{P_0}^{P_e} B^n(p)\, dP \right)^{\frac{1}{n}} \qquad (4.53)$$

where l is the length of the porous specimen.

Due to the flow rate of a power-law liquid

$$v_0 = B_0 \left(\frac{\Delta P}{l}\right)^{\frac{1}{n}}$$

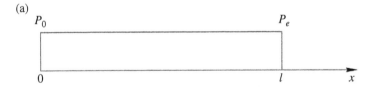

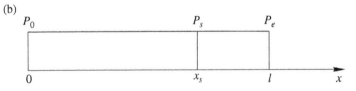

Figure 4.22 Diagram of one-dimensional flow flux in the porous cylindrical specimen: (a) $P_S > P_0 > P_e > P_c$ and (b) $P_0 > P_S > P_e > P_c$.

the ratio of the liquid flow rates will be

$$\frac{v}{v_0} = \frac{Q}{Q_0} = \frac{1}{B_0}\left(\frac{-\int_{P_0}^{P_e} B^n(p)\,dP}{\Delta P}\right)^{\frac{1}{n}} \quad (4.54)$$

In order to determine the effect of the slip on flow, the calculations for a porous medium with a permeability of 0.1 μm² and a porosity of 0.25 using formulas (4.53) and (4.54) at different values of n (the integral was solved numerically because it is taken analytically only for even values of n) were conducted. The values of constants k_0 were defined at $\tau = 2\,\text{N m}^{-2}$ and $\gamma = 1312\,\text{s}^{-1}$. The values of other parameters were taken from reference [34]. The calculations were performed using the constant pressure P_0 at the inlet of the porous medium.

The results of these calculations are shown in Figure 4.23. It is seen that the pattern of flow is not changed qualitatively for initially pseudoplastic liquids (Figure 4.23a, b), but the flow rate rises significantly at the constant pressure. For the originally dilatant liquids, the initial flow curves are changed completely (Figure 4.23c, d). For example, at $n = 1.1$, the dilatant liquid changes the pattern of flow to pseudoplastic (Figure 4.23c). At $n = 3$, the pattern changes to S-shaped (Figure 4.23d). As seen from Figures 4.24 and 4.25, the dependences of the ratio of flow rates on pressure drop exhibit an S-shaped character at both $n < 1$ (Figure 4.24) and $n > 1$ (Figure 4.25).

Let us consider the more general case with $P_0 > P_S > P_e > P_c$, when the porous medium has regions with and without slip (Figure 4.22b). Solving Equations (4.50) and (4.52), it can be found that

$$B(p) = \left(\frac{dP}{dx}\right)^{\frac{1}{n}} = v_S$$

where v_S is the flow rate at the point x_S. The pressure distribution is described as follows:

$$-\int_{P}^{P_0} B^n(p)\,dP = v_S^n \int_{x}^{0} dx$$

For the region $0 < x < x_S$, where $B(p) = B_0$, i.e. for the region without slip, the following equation can be obtained:

$$-\int_{P_S}^{P_0} dP = \left(\frac{v_S}{B_0}\right)^n \int_{x_S}^{0} dx \quad (4.55)$$

For the region with slip $x_S < x < l$,

$$-\int_{P_e}^{P_S} B^n(p)\,dP = v_S^n \int_{l}^{x_S} dx$$

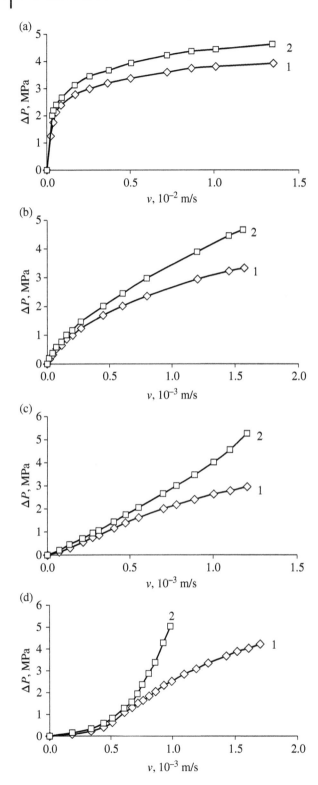

Figure 4.23 Calculated flow curves (1) with slip and (2) without lip at: (a) n = 0.2; (b) n = 0.7; (c) n = 1.1, and (d) n = 3.

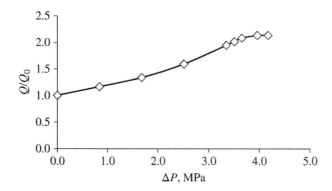

Figure 4.24 Dependence of the ratio of liquid flow rates (Q/Q_0) on a pressure drop at $n = 0.7$.

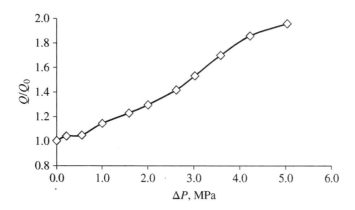

Figure 4.25 Dependence of the ratio of liquid flow rates (Q/Q_0) on a pressure drop at $n = 1.1$.

After some transformations, the contact point of the two regions is found:

$$x_S = l\left[1 + \frac{\int_{P_e}^{P_S} B^n(p)\,dP}{B_0^n(P_0 - P_S)}\right]^{-1}$$

The expression for the flow rate with slip is obtained from Equation (4.55):

$$v = B_0\left(\frac{P_0 - P_S}{x_S}\right)^{\frac{1}{n}}$$

4 Nanogas Emulsions in Oil Field Development

Hence, the ratio of liquid flow rates is expressed as follows:

$$\frac{v}{v_0} = \frac{Q}{Q_0} = \left(\frac{P_0 - P_S}{P_0 - P_e} \frac{l}{x_S}\right)^{\frac{1}{n}}$$

Pressure distributions $\frac{v}{v_0} - \frac{Q}{Q_0} = \left(\frac{P_0 - P_S}{P_0 - P_e} \frac{l}{x_S}\right)^{\frac{1}{n}}$ calculated at $P_e = 4$ MPa, $P_s = 8$ MPa, and $P_0 = 16$ MPa for both initially pseudoplastic ($n = 0.7$) and dilatant ($n = 3$) liquids are shown in Figure 4.26. It can be seen that the slip effect in both cases results in a significant decrease in the slope of pressure distribution curves within the $P_S = P_e$ interval. This is responsible for an increase in the liquid flow rate and the variation of rheological characteristics.

It can be seen that the variation in the initial rheological characteristics is possible during the flow of gasified non-Newtonian liquids in the subcritical region.

4.2.2.3 Flow in a Heterogeneous Porous Medium

A simplified scheme of non-Newtonian liquid flow in a nonhomogeneous stratified bed has been analyzed previously [35]. In line with this, the flow occurs in two parallel layers with permeabilities of k_1 and k_2 (other conditions being equal); when $k_1 \gg k_2$ the liquid obeys the power law. Then, according to formula (4.50), at $B = B_0$ the following equality is obtained:

$$Q_1 Q_2 = (k_1/k_2)(k_1/k_2)^{(1-n)/2n}$$

It can be seen that the ratio of flow rates is equal to the ratio of permeabilities in the case of a Newtonian liquid ($n = 1$), it decreases for a dilatant liquid ($n > 1$), i.e. the flow profile equalizes, and, finally, the difference in permeability increases for a pseudoplastic liquid ($n < 1$).

According to the obtained results, the joint action of dilatancy and the slip effect must favor an equalization of the flow profile. Indeed, let us now assume that the flow of the gasified non-Newtonian liquid under a near-transition phase state occurs in two parallel layers

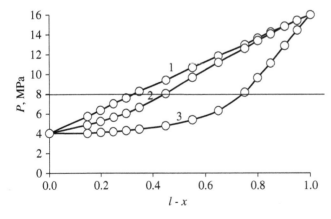

Figure 4.26 Pressure distribution in the case of $P_0 > P_S > P_e > P_c$ (1) without slip, (2) with slip for pseudoplastic liquid ($n = 0.7$), and (3) with slip for dilatant liquid ($n = 3$).

with permeabilities k_1 and k_2 (other conditions being equal), $k_1 \gg k_2$, and corresponds to the power law with slip. Then, according to Equation (4.50), the following equation is obtained:

$$\frac{Q_1}{Q_2} = \frac{k_1}{k_2}\left(\frac{k_1}{k_2}\right)^{\frac{1-n}{2n}}\left(1+\frac{3n+1}{n}\frac{b\sqrt{m}}{\sqrt{8k_1}}\right) \times \left(1+\frac{3n+1}{n}\frac{b\sqrt{m}}{\sqrt{8k_2}}\right)^{-1}$$

As seen from the last expression, the slip enhances the effect of equalization of the flow profile in the case of a dilatant liquid ($n > 1$) and neutralizes the negative effect of shear fluidization on the flow profile for a pseudoplastic liquid ($n < 1$).

Note that the bubbles of solutes in a liquid evolved at the surface of the porous medium during flow of the gassed power-law liquid prevent adsorption and accelerate the flow.

4.2.2.4 Concluding Remarks

1) Variations in the initial rheological characteristics are possible during flow of gasified non-Newtonian (power-law obeying) liquids in a subcritical region.
2) The slip during the flow through a nonhomogeneous porous medium enhances the equalization of the flow profile in the case of a dilatant liquid and neutralizes the negative effect of shear fluidization on the flow profile in the case of a pseudoplastic liquid.
3) The relationships proposed may be used to analyze the flow of non-Newtonian liquids with slip.

4.3 Field Validation of Slippage Phenomena

4.3.1 Steady-State Radial Flow

4.3.1.1 Gasified Newtonian Fluid Flow

For this analysis let us consider the steady-state radial flow of a quasi-homogeneous incompressible fluid. The slippage effect will be taken into account and the overall process is defined by the following equations (unless otherwise stated, the notations adopted in Section 2.4.1 are used):

$$\frac{d}{dr}(rv) = 0$$

$$v = \frac{k(P)}{\eta}\frac{dP}{dr}$$

(4.56)

where $k(P)$ is the effective permeability.

Consider the case when $P_s \geq P_k > P_e \geq P_c$, where P_k and P_e are the drainage area and bottomhole pressures respectively (see the diagram in Figure 4.27a) and the slippage effect occurs across the whole flow region.

Solution for the fluid flow rate at the (4.56) system could be expressed as:

$$Q = \frac{2\pi h k(P)}{\eta}\frac{dP}{dr}r \qquad (4.57)$$

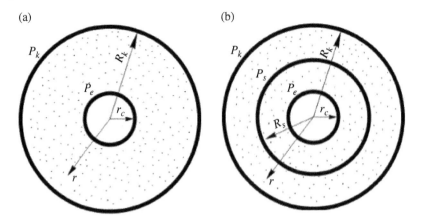

Figure 4.27 Radial flow scheme: (a) $P_s \geq P_k > P_e \geq P_c$; (b) $P_k > P_s > P_e \geq P_c$.

where h is the formation thickness (overall flow distance).
From Equation (4.57) for the whole flow, one gets

$$Q = \frac{2\pi h}{\eta \ln \frac{R_k}{r_c}} \int_{P_e}^{P_k} k(P)\,dP$$

where R_k and r_c are the drainage area radius and well radius respectively.
The pressure distribution term is resolved as

$$\int_P^{P_k} k(P)\,dP = \frac{\ln \frac{R_k P_k}{r}}{\ln \frac{R_k}{r_c}} \int_{P_e}^{P_k} k(P)\,dP$$

The Dupuit flow rate is determined by the equation

$$Q_0 = \frac{2\pi k_0 h(P_k - P_e)}{\eta \ln \frac{R_k}{r_c}}$$

For the fluid flow rate $Q_1 = Q / Q_0$,

$$Q_1 = \frac{\int_{P_e}^{P_k} k(P)\,dP}{k_0(P_k - P_e)}$$

The specified expression, in general terms, is identical to the same expression for the straight-line flow (4.31), so the dependencies, obtained on this basis, are also valid for the radial case.

The obtained results are also in good agreement with the field data presented in reference [36]. Figures 4.28 and 4.29 show the data from Figures 2.2 and 2.10 of that mentioned work. It can be seen that the field data have good agreement with the proposed model both qualitatively and quantitatively.

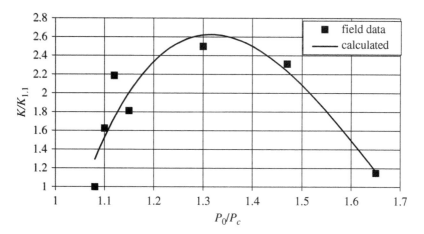

Figure 4.28 Dependence of the ratio of the permeability of matrix blocks ($K_{1.1}$ – permeability at $P_0/P_c = 1.1$) on P_0/P_c for well X0 of the Bach Ho (White Tiger) oil field [36].

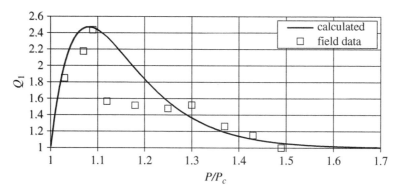

Figure 4.29 Dependence of the dimensionless flow rate (Q_0 is taken equal to the flow rate $P_0/P_c = 1.7$) on P/P_c for the Bach Ho (White Tiger) oil field [36].

In the case where $P_k > P_s > P_e > P_c$ (see the diagram in Figure 4.27b), there are two regions: flow without slippage phenomena in the region $P_s < P < P_k$, i.e. $k(P) = K_0$, and flow with slippage phenomena in the region $P_e \leq P < P_s$.

The pressure distribution for the first region can be determined from the following equation:

$$P = P_k - (P_k - P_s)\frac{\ln\dfrac{R_k}{r}}{\ln\dfrac{R_k}{R_s}}$$

For the second region, it can be obtained from

$$\int_P^{P_s} k(P)dP = \frac{\ln\dfrac{R_s\,P_s}{r}}{\ln\dfrac{R_s}{r_c}P_e}\int_{P_e}^{P_s} k(P)dP \tag{4.58}$$

In order to determine $\ln R_s$ and taking into account the continuity of motion (see Equation (4.56)) the expressions (4.59) and (4.60) have been obtained for the first and second regions respectively:

$$Q = \frac{2\pi k_0 h (P_k - P_s)}{\eta \ln \frac{R_k}{R_s}} \qquad (4.59)$$

$$Q = \frac{2\pi h}{\eta \ln \frac{R_s}{r_c}} \int_{P_e}^{P_s} k(P)\,dP \qquad (4.60)$$

The ratio of Equations (4.59) and (4.60) then gives

$$\frac{\ln R_s}{\ln r_c} = \frac{\left(1 + \frac{\ln R_k}{\ln r_c} \frac{\int_{P_e}^{P_s} k(p)\,dp}{k_0(P_k - P_s)}\right)}{\left(1 + \frac{\int_{P_w}^{P_s} k(p)\,dp}{k_0(P_k - P_s)}\right)}$$

and for the fluid flow rate one can obtain

$$Q = \frac{2\pi k_0 h (P_k - P_s)}{\eta \ln \frac{R_k}{R_s}}$$

For the dimensionless fluid flow rate,

$$Q_1 = \frac{(P_k - P_s) \ln \frac{R_k}{r_c}}{(P_k - P_3) \ln \frac{R_k}{R_s}} \qquad (4.61)$$

Expansion of the integral $\int_{P_1}^{P_2} k(P)\,dP$ is given in Section 4.1.2.3.

Figure 4.30 shows the $\Delta P(Q)$ calculation dependencies (for the case $P_s \geq P_k > P_e \geq P_c$) at the following values of the flow parameters: $k_0 = 20$ mD, $h = 10$ m, $R_k = 25$ m, $r_c = 0.15$ m, and $\eta = 5$ mP; the other parameters were taken from Section 4.1. It can be seen from the figure that the obtained results correspond to the same results for linear flow (See Section 4.1). Here, also at different scenarios of pressure differential alteration, both the mainly pseudoplastic (at a constant drainage area pressure) and the mainly dilatant (at a

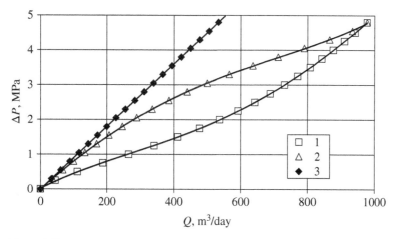

Figure 4.30 Calculated pressure–flow rate dependences: 1 – with slippage at P_e – constant, 2 – with slippage at P_k – constant, 3 – according to Dupuit.

constant bottomhole pressure) flows are possible. Furthermore, in both cases, the fluid flow rate is higher than the Dupuit flow rate at a constant pressure differential. It should be noted that the first scenario is carried out in field conditions since the drainage area pressure change is a capital-intensive operation, so the pseudoplastic nature of the pressure–flow rate dependences should be expected in field conditions in the presence of slippage.

The obtained results agree well with the field data, presented in reference [36]. Figures 4.31 and 4.32 presented the data of Figure 1.17 from the mentioned work. It can be seen from the figures that the given pressure–flow rate dependences taken at the pressure near (above) the saturation pressure are indicative of the pseudoplastic flow nature.

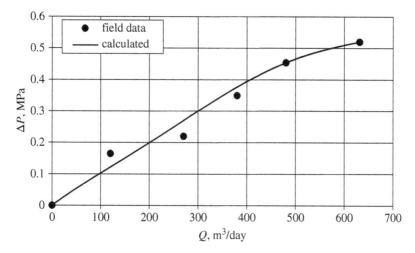

Figure 4.31 Pressure–flow rate dependence for well B03 of the Bach Ho (White Tiger) oil field [36].

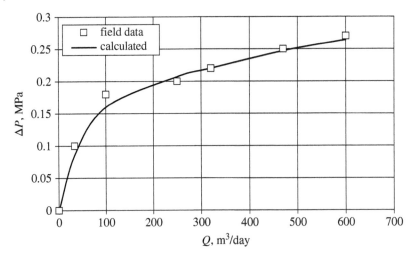

Figure 4.32 Pressure–flow rate dependence for well B38 of the Bach Ho (White Tiger) oil field [36].

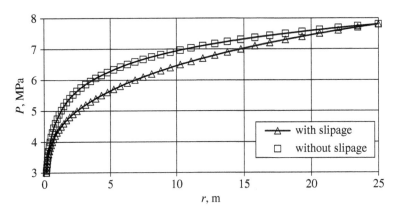

Figure 4.33 Pressure distribution with and without the slippage effect.

Figure 4.33 shows the pressure distribution for the case when $P_s \geq P_k > P_e \geq P_c$ and in the absence of slippage. It can be seen from the figure that the presence of slippage leads to a reduction in the pressure at the same radius of the radial flow, which conditions the fluid flow rate growth.

Let us examine the case when the bottomhole pressure is below the saturation pressure, i.e. $P_s \geq P_k > P_c > P_e$ (see the diagram in Figure 4.27b and take P_c and R_c instead of P_s and R_s respectively). In this case, there are also two regions: flow with slippage occurs in the region $P_c < P < P_k$ and two-phase system flow takes place in the region $P_e < P < P_c$. As shown above, in the second region, the flow can be represented as a homogeneous incompressible fluid movement. The effective permeability for the later can be taken in the form

$$k(P) = k_0 \left(\frac{P}{P_c}\right)^n$$

where $n = 0.42$.

Based on Equation (4.58), the pressure distribution for the first region can be determined from the expression

$$\int_P^{P_k} k(P)\, dP = \frac{\ln \dfrac{R_k P_k}{r}}{\ln \dfrac{R_k}{R_c} P_c} \int k(P)\, dP$$

For the second region:

$$P^{n+1} = P_c^{n+1} - \left(P_c^{n+1} - P_e^{n+1}\right) \frac{\ln \dfrac{R_c}{r}}{\ln \dfrac{R_c}{r_c}}$$

In order to determine $\ln R_c$ and by taking into account the continuity of flow, for the first region the flow can be described by the following expression:

$$Q = \frac{2\pi h}{\eta \ln \dfrac{R_k}{R_c} P_c} \int_{}^{P_k} k(P)\, dP \tag{4.62}$$

For the second region,

$$Q = \frac{2\pi k_0 h}{\eta \ln \dfrac{R_c}{r_c}} \frac{\left(P_c^{n+1} - P_e^{n+1}\right)}{(n+1) P_c^n} \tag{4.63}$$

Combining into the ratio:

$$\frac{\ln R_c}{\ln r_c} = \frac{\left(1 + \dfrac{\ln R_k}{\ln r_c} \dfrac{k_0 \left(P_c^{n+1} - P_e^{n+1}\right)}{(n+1) P_c^n \displaystyle\int_{P_c}^{P_k} k(p)\, dp}\right)}{\left(1 + \dfrac{k_0 \left(P_c^{n+1} - P_e^{n+1}\right)}{(n+1) P_c^n \displaystyle\int_{P_c}^{P_k} k(p)\, dp}\right)}$$

Finally, for the fluid flow rate,

$$Q = \frac{2\pi h}{\eta \ln \dfrac{R_k}{R_c} P_c} \int_{}^{P_k} k(P)\, dP$$

and the dimensionless flow rate is expressed as

$$Q_1 = \frac{\left(\displaystyle\int_{P_c}^{P_k} k(p)\, dp\right) \ln \dfrac{R_k}{r_c}}{k_0 (P_k - P_e) \ln \dfrac{R_k}{R_c}} \tag{4.64}$$

4 Nanogas Emulsions in Oil Field Development

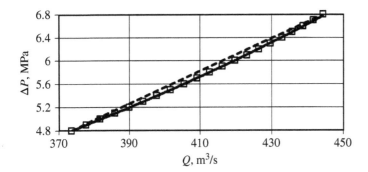

Figure 4.34 Calculated pressure–flow rate dependence for case $P_s \geq P_k > P_e \geq P_c$.

Figure 4.34 shows the pressure–flow rate dependence, calculated according to the formula mentioned above (at P_e – constant). It can be seen from the figure that the flow nature, as opposed to flow with slippage, is dilatant.

The pressure distribution for a case with $P_e = 1$ MPa is shown in Figure 4.35. It can be seen that, while the bottomhole pressure is higher than the saturation pressure, the pressure at the same radius of the radial flow is lower (flow resistance is reduced) and, at the bottomhole pressure below the saturation pressure, it is higher than the Dupuit pressure (flow resistance is increased).

Figure 4.36 shows the calculated dependence of the dimensionless fluid flow rate on the above discussed pressure level. It can be seen from the figure that, where the bottomhole pressure is below the saturation pressure, the dimensionless flow rate decreases sharply and becomes smaller than one, despite the fact that the drainage area pressure is higher than the saturation pressure.

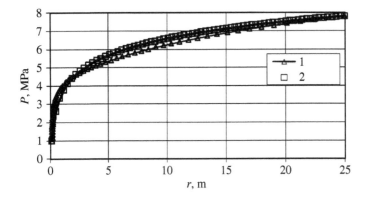

Figure 4.35 Pressure distribution for case $P_s \geq P_k > P_e \geq P_c(1)$ and according to Dupuit (2).

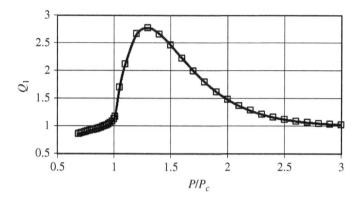

Figure 4.36 Dependence of the dimensionless flow rate on the pressure level.

4.3.1.2 Gasified Non-Newtonian Fluid Flow

In a case where the Kozeny-Carman porous medium occurs, the following equation can be used:

$$Q = 2\pi h B(P) r \left(\frac{dp}{dr}\right)^{\frac{1}{n}} \tag{4.65}$$

Let us consider a steady-state radial flow for the case $P_s \geq P_k > P_e \geq P_c$, i.e. slippage occurs throughout the fluid flow (see the diagram in Figure 4.27a). Solving Equation (4.65) together with the continuity equation for an incompressible fluid:

$$\frac{d}{dr}(rv) = 0 \tag{4.66}$$

it is possible to get

$$Q = 2\pi h \left(\frac{1-n}{R_k^{1-n} - r_c^{1-n}} \int_{P_e}^{P_k} B^n(P)\, dP \right)^{\frac{1}{n}} \tag{4.67}$$

The pressure distribution can be calculated from the following:

$$\int_P^{P_k} B^n(P)\, dP = \frac{R_k^{1-n} - r^{1-n}}{R_k^{1-n} - r_c^{1-n}} \int_{P_e}^{P_k} B^n(P)\, dP$$

The dependences define the flow rate of the exponential law fluid:

$$Q = 2\pi h B_0 \left(\frac{(1-n)(P_k - P_e)}{R_k^{1-n} - r_c^{1-n}} \right)^{\frac{1}{n}}$$

and for the flow rates ratio:

$$Q_1 = \frac{1}{B_0} \left(\frac{\int_{P_a}^{P_k} B^n(p)\, dP}{(P_k - P_e)} \right)^{\frac{1}{n}} \tag{4.68}$$

This is identical to the same expression for the linear flow (see Section 4.1) so the dependencies obtained on its basis are also valid for the radial case.

In order to determine the slippage influence on the nature of the pressure–flow rate dependencies, the calculations were performed in accordance with the equations above at different n values for the porous medium with the permeability of 0.02 μm² and porosity of 0.2 (the integral was determined numerically, as it only expands to an even n). It was found at $\tau = 2\,\mathrm{N/m^2}$, $\gamma = 400\,\mathrm{s^{-1}}$. The other parameters were taken from the paragraph "a." The results are shown in Figures 4.37–4.40. It can be seen from the figures that the obtained results are basically the same as the results for the straight-line flow. Indeed, at P_k constant for the initially pseudoplastic fluids (Figures 4.37 and 4.38), no qualitative flow changes occur, although a significant fluid flow rate increase is observed at a constant pressure differential. For the initially dilatant fluids, a complete modification of the flow nature takes place (Figures 4.39 and 4.40). Thus, at $n = 1.1$, the dilatant fluid changes the nature of flow into a pseudoplastic one (Figure 4.39) and, at $n = 3$, into the S-shaped one (Figure 4.40).

At P_e constant for the initially dilatant fluids (Figures 4.39 and 4.40), no qualitative flow changes occurred, though there was a significant fluid flow rate increase at the constant pressure differential. For the initially pseudoplastic fluids, complete modification of the flow nature takes place (except for a small initial section) at certain values of n (Figure 4.38). It should be noted that the $\Delta P(Q)$ dependence peculiarities mentioned in Section 4.2.2 for the straight-line flow (combined nature with the predominance of the

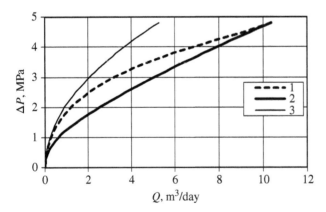

Figure 4.37 Calculated pressure–flow rate dependences at $n = 0.5$: (1) P_κ – constant; (2) P_e – constant; (3) exponential law fluid.

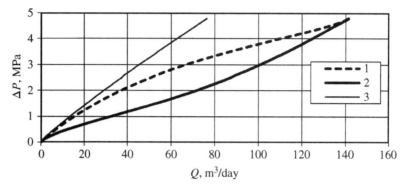

Figure 4.38 Calculated pressure–flow rate dependences at $n = 0.9$: (1) P_κ – constant; (2) P_e – constant; (3) exponential law fluid.

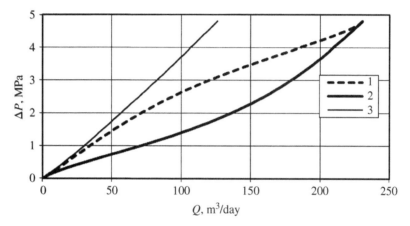

Figure 4.39 Calculated pressure–flow rate dependences at $n = 1.1$: (1) P_κ – constant; (2) P_e – constant; (3) exponential law fluid.

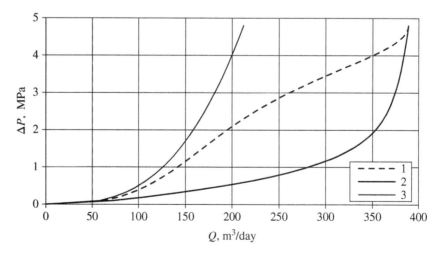

Figure 4.40 Calculated pressure–flow rate dependences at $n = 3$: (1) P_κ – constant; (2) P_e – constant; (3) exponential law fluid.

pseudoplastic or dilatant flow nature of the $Q_l(\Delta)P$ dependence) are also true for this case, which can be clearly seen from the figures shown above.

The obtained results agree well with the field data presented in reference [36] (at P_κ constant). Figures 4.41 and 4.42 show the data of Figures 2.7 and 2.9 from the mentioned work. It can be seen from Figure 4.41 that when the slippage effect strengthens and the pressure approaches the saturation pressure, the fluid flow rate significantly increases at other equal conditions. It can be seen from Figure 4.42 that, when the slippage effect strengthens, the flow nature changes qualitatively from the dilatant flow into the S-shaped one.

Figure 4.43 shows the pressure distribution at different n values. It can be seen from the figure that the presence of slippage results in a pressure decrease at other equal conditions.

Now we consider a more general case $P_k > P_s > P_e \geq P_c$ (see the diagram in Figure 4.27b). In this case, there are also two regions: flow without slippage occurs in the region $P_s < P < P_k$, i.e. $B(P) = B_0$, and flow with slippage occurs in the region $P_e \leq P < P_s$.

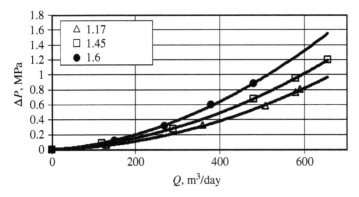

Figure 4.41 Pressure–flow rate dependences of well X0 the Bach Ho (White Tiger) oil field [36] at different P/P_c.

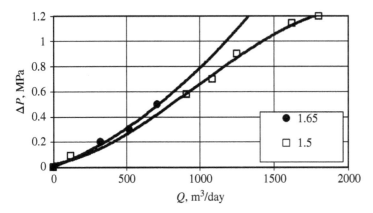

Figure 4.42 Pressure–flow rate dependences of well B03 the Bach Ho (White Tiger) oil field [36] at different P/P_c.

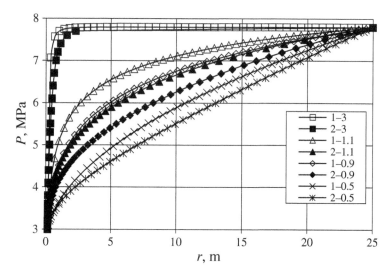

Figure 4.43 Pressure distribution at different n. (1) without slippage; (2) with slippage.

When solving Equations (4.65) and (4.66) for the pressure distribution in the first region, the following equation can be obtained:

$$P = P_k - (P_k - P_s)\frac{R_k^{1-n} - r^{1-n}}{R_k^{1-n} - R_s^{1-n}}$$

The second region is defined as follows:

$$\int_P^{P_s} B^n(P)\,dP = \frac{R_s^{1-n} - r^{1-n}}{R_s^{1-n} - r_c^{1-n}} \int_{P_e}^{P_s} B^n(P)\,dP$$

In order to determine R_s and by taking into account the continuity of motion for the first region, the flow is defined by

$$Q = 2\pi h B_0 \left(\frac{(1-n)(P_k - P_s)}{R_k^{1-n} - R_s^{1-n}}\right)^{\frac{1}{n}} \tag{4.69}$$

and for the second region

$$Q = 2\pi h \left(\frac{1-n}{R_s^{1-n} - r_c^{1-n}} \int_{P_e}^{P_s} B^n(P)\,dP\right)^{\frac{1}{n}} \tag{4.70}$$

Finally, the ratio is

$$R_s^{1-n} = \frac{r_c^{1-n}\left(1 + \dfrac{R_k^{1-n}}{r_c^{1-n}} \dfrac{\int_{P_e}^{P_s} B^n(p)\,dp}{B_0^n(P_k - P_s)}\right)}{\left(1 + \dfrac{\int_{P_e}^{P_s} B^n(p)\,dp}{B_0^n(P_k - P_s)}\right)}$$

From Equation (4.69) or Equation (4.70) for the fluid flow rate:

$$Q = 2\pi h B_0 \left(\frac{(1-n)(P_k - P_s)}{R_k^{1-n} - R_s^{1-n}}\right)^{\frac{1}{n}}$$

and for the flow ratio the following dependence was obtained:

$$Q_1 = \left(\frac{(P_k - P_s)}{(P_k - P_e)} \frac{(R_k^{1-n} - r_c^{1-n})}{(R_k^{1-n} - R_s^{1-n})}\right)$$

In conclusion, the calculated results for the radial case are in good agreement with the field studies for the gassed fluid flow in the subcritical region.

4.3.2 Unsteady State Flow

As the mentioned experimental studies showed, the slippage effect exerts a significant influence on the process of pressure build-up during the gassed fluid flow. Hence, the study of the pressure build-up with the account of the slippage effect at nonimmediate stoppage of the fluid inflow is both scientifically and practically interesting. For this study it was considered as a model consisting of a well with the initial flow rate Q_0 and a fluid with some inertia (some flow when the pressure differential is removed).

It is a simplified model with an assumption that the porous medium permeability depends on the pressure (unless otherwise stated, the notations adopted in Sections 2.4.1 and 2.4.2 are used). The pressure distribution in the formation is described by the equation of piezo-conductivity:

$$\frac{\partial P}{\partial t} = \frac{1}{r}\frac{\partial}{\partial r}\left[\chi(P)r\frac{\partial P}{\partial r}\right] \tag{4.71}$$

The initial and boundary conditions are

$$P|_{t=0} = P_{cm} \tag{4.72}$$

$$P|_{r=R_k} = P_k 2\pi rh \frac{k(P)}{\eta} \frac{\partial p}{\partial r}|_{r=r_c} = Q(t)\} \tag{4.73}$$

In order to solve the nonlinear boundary-value problem of Equation (4.71), the Kirchhoff transformation was used

$$\theta = \frac{1}{\chi_0} \int_0^P \chi(P) \, dP \tag{4.74}$$

From Equation (4.71), with the help of Equation (4.74),

$$\frac{\partial \theta}{\partial t} = \chi_0 \frac{1}{r} \frac{\partial}{\partial r} \left[r \frac{\partial \theta}{\partial r} \right] \tag{4.75}$$

Averaging $\partial \theta/\partial t$ over r gives

$$\phi(t) = \frac{2}{R_k^2 - r_c^2} \int_{r_c}^{R_k} \frac{\partial \theta}{\partial t} r \, dr \tag{4.76}$$

It is shown in Section 2.4.1 that the effective permeability with the account of the fluid slippage can be determined by the expression

$$k(P) = k_0 \left[1 + \frac{4b(P)}{R} \right] \tag{4.77}$$

The assumption is that the flow process happens at above the saturation pressure when the slippage effect is taken into account.

When linearizing the results obtained in reference [34] and shown in Section 2.4.1, for the piezo-conductivity,

$$\left.\begin{aligned} \chi &= \frac{k_0}{\eta \beta^*} \left[1 + \frac{4}{R} a(P - P_c) \right] & P_c \leq P \leq P_m \\ \chi &= \frac{k_0}{\eta \beta^*} \left[1 + \frac{1}{R} a(P_s - P) \right] & P_m \leq P \leq P_s \end{aligned}\right\} \tag{4.78}$$

where P_m is the pressure at which the effective permeability reaches its maximum value (in experiments $P_m = 1.1 - 1.5 P_c$) and a is a constant.

From Equation (4.74), with the use of Equation (4.78),

$$\left.\begin{aligned} \theta &= P + \frac{2a}{R} \left[(P - P_c)^2 - P_c^2 \right] & P_c \leq P \leq P_m \\ \theta &= P + \frac{a}{2R} \left[P_s^2 - (P_s - P)^2 \right] & P_m \leq P \leq P_s \end{aligned}\right\} \tag{4.79}$$

The pressure is presented in the form

$$P = P_{st} + P_f \tag{4.80}$$

where

$$P_{st} = P_\kappa + \frac{\eta Q_0}{2\pi k_0 h} \ln \frac{r}{R_\kappa}$$

is the pressure stationary element and P_f is the pressure fluctuating element.

When inserting Equation (4.80) into Equation (4.79) and linearizing, the following expressions were obtained:

$$\theta = P_{st} + P_f + \frac{2a}{R}\left[\left(1 + \frac{2P_f}{P_{st}}\right)P_{st}^2 - 2P_{st}P_c - 2P_cP_f\right] \qquad P_c \leq P \leq P_m \qquad (4.81)$$

$$\theta = P_{st} + P_f + \frac{a}{2R}\left[-\left(1 + \frac{2P_f}{P_{st}}\right)P_{st}^2 + 2P_sP_{st} + 2P_sP_f\right] \qquad P_m \leq P \leq P_s$$

The boundary condition (4.73), taking account of Equations (4.77) and (4.79), takes the form

$$\frac{k(P)}{\eta}2\pi h r_c \left.\frac{\partial P}{\partial r}\right|_{r=r_c} = \frac{k_0}{\eta}2\pi h r_c \left.\frac{\partial \theta}{\partial r}\right|_{r=r_c} \qquad (4.82)$$

After inserting Equation (4.76) into Equation (4.75) and integrating it with Equations (4.72), (4.73), and (4.82), the following equation was obtained:

$$\theta = \theta_k + \frac{\varphi(t)}{2\chi_0}(r^2 - R_k^2) - \frac{\varphi(t)}{2\chi_0}r_c^2 \ln\left(\frac{r}{R_k}\right) + \frac{\mu Q(t)}{2\pi h k_0}\ln\frac{r}{R_k} \qquad (4.83)$$

where

$$\theta_k = P_k + \frac{2a}{R}\left[(P_k - P_c)^2 - P_c^2\right], \qquad P_c \leq P \leq P_m$$

$$\theta_k = P_k + \frac{a}{2R}\left[P_s^2 - (P_s - P_k)^2\right], \qquad P_m \leq P \leq P_s$$

Then, from Equation (4.76) together with Equation (4.83) and neglecting r_c^2, we can get

$$\dot{\varphi} + \frac{4\chi_0}{R_k^2}\varphi = -\frac{\dot{Q}(t)}{\pi\beta^* h R_k^2} \qquad (4.84)$$

where $\dot{\varphi}\dot{Q}(t)$ are time derivatives of φ and $Q(t)$.

It should be noted that, at a constant value of formation permeability k_0, the differential equation for φ will have the same form as Equation (4.84).

Flow time dependence should be determined from the simultaneous solution of the differential equations of flow and fluid momentum in the vertical pipeline. However, in the first approximation, it is possible to think that, when the fluid flow is instantly cut off from the wellhead, the well inflow immediately stops but does not change according to the law

$$Q = Q_0 e^{-\beta_0 t} \qquad (4.85)$$

where $\beta_0 = 4\chi_0/R_k^2$.

Then, from Equation (4.84) and taking account of Equation (4.85), at a pressure much above the saturation pressure and when there is no fluid slippage ($P > P_s$), the following equation was obtained:

$$\phi = \frac{Q_0}{\pi\beta^* h R_k^2}\exp\left(-\frac{4\chi_0}{R_k^2}t\right)\left(\frac{4\chi_0}{R_k^2}t - 1\right) + C_1\exp\left(-\frac{4\chi_0}{R_k^2}t\right) \qquad (4.86)$$

4.3 Field Validation of Slippage Phenomena

From Equation (4.76) and taking account of Equations (4.72) and (4.86), the pressure distribution was expressed as follows:

$$P = P_k + \frac{r^2 - R_k^2}{2\chi_0} \frac{Q_0}{\pi \beta^* h R_k^2} \exp\left(-\frac{4\chi_0}{R_k^2} t\right) \frac{4\chi_0}{R_k^2} t + \frac{\eta Q_0 \exp\left(-\frac{4\chi_0}{R_k^2} t\right)}{2\pi h k_0} \ln\frac{r}{R_k} \quad (4.87)$$

The time change (build-up) of the bottomhole pressure can be determined from Equation (4.87) and will have the form

$$P_e = P_k - \frac{2Q_0 t}{\pi \beta^* h R_k^2} \exp\left(-\frac{4\chi_0}{R_k^2} t\right) + \frac{\mu Q_0 \exp\left(-\frac{4\chi_0}{R_k^2} t\right)}{2\pi h k_0} \ln\frac{r_c}{R_k} \quad (4.88)$$

or

$$P_e = P_k - \Delta P \exp\left(-\frac{4\chi_0 t}{R_k^2}\right)\left(\frac{4\chi_0 t}{R_k^2 \ln\frac{R_k}{r_c}} + 1\right)$$

Pressure build-up at the presence of fluid slippage can be determined from Equation (4.76) and taking account of Equations (4.72), (4.79), and (4.86), based on the following formulas:

at $P_c \leq P \leq P_m$

$$P_e = P_k - \frac{R}{4a}\left[1 + \frac{4a}{R}(P_k - P_c)\right] +$$

$$\sqrt{\left(\frac{R}{4a}\right)^2 \left[1 + \frac{4a}{R}(P_k - P_c)\right]^2 - \frac{R}{2a}\Delta P \exp\left(-\frac{4\chi_0 t}{R_k^2}\right)\left[\frac{4\chi_0 t}{R_k^2 \ln\left(\frac{R_k}{R_c}\right)} + \frac{2a}{R}(P_{e0} + P_k - 2P_c) - 1\right]}$$

at $P_m \leq P \leq P_s$

$$P_e = P_k + \frac{R}{a}\left[1 + \frac{a}{R}(P_s - P_k) - \frac{a}{2R}\Delta P \exp\left(-\frac{4\chi_0}{R_k^2} t\right)\right] -$$

$$\sqrt{\left(\frac{R}{a}\right)^2 \left[1 + \frac{a}{R}(P_s - P_k) - \frac{a}{2R}\Delta P \exp\left(-\frac{4\chi_0}{R_k^2} t\right)\right]^2 + 2\Delta P \exp\left(-\frac{4\chi_0 t}{R_k^2}\right)\left[P_s - P_k + \frac{R}{a}\left(\frac{4\chi_0 t}{R_k^2 \ln\frac{R_k}{r_c}} + 1\right)\right]}$$

where P_{e0} is the initial bottomhole pressure, $\Delta P = P_k - P_{e0}$.

Figures 4.44–4.47 show the results of the numerical calculation on the basis of the produced equations at the following values of the parameters: $P_k = 10.8$–22 MPa; $\eta = 1$ mPa s; $h = 10$ m; $r_c = 0.15$ m; $\chi_0 = 0.05$ m² s^{-1}; $k_0 = 0.15 \times 10^{-12}$ m²; $m = 0.2$; $a = 5 \times 10^{-11}$ m^{-1}; $R = 2.45 \times 10^{-6}$ m; $\beta^* = 2.94 \times 10^{-9}$ 1 Pa^{-1}; $R_\kappa = 25$ m; $P_s = 20$ MPa; $P_c = 8.0$ MPa; $P_{e0} = 8.8$–20 MPa.

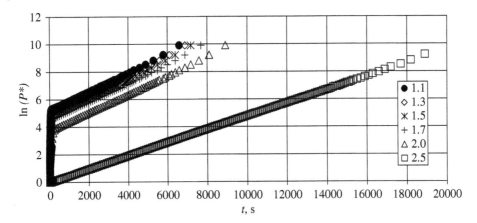

Figure 4.44 Calculated pressure build-up curves, replotted in semi-logarithmic coordinates at different P/P_c.

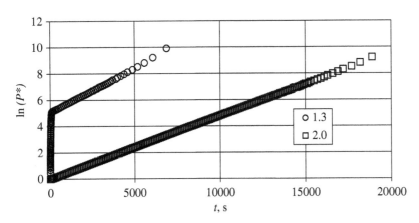

Figure 4.45 Calculated pressure build-up curves, replotted in semi-logarithmic coordinates at different P/P_c.

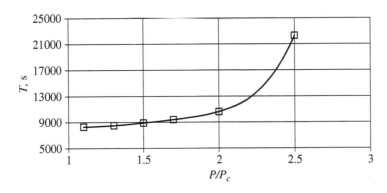

Figure 4.46 Dependence of pressure build-up time on P/P_c.

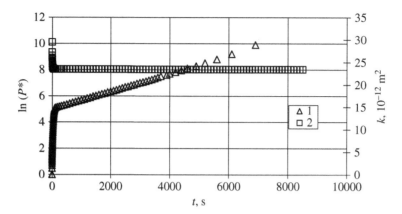

Figure 4.47 Calculated pressure build-up curve, replotted in semi-logarithmic coordinates (1) and dynamics of porous medium permeability (2) at $P/P_c = 1.3$.

Calculated pressure build-up curves are replotted in semi-logarithmic coordinates. It can be seen from Figures 4.44 and 4.45, at $P < 2P_c$, that the obtained dependence has the form of a polygonal line with two characteristic sections and, at $P \geq 2.5 P_c$, it has the form of a straight line, i.e. when the initial bottomhole pressure approximates P_c, the exponential model changes into the model described by the sum of exponents. When approaching the critical region, the curves deviate to the left (i.e. toward smaller relaxation times); moreover, the pressure build-up time also decreases (Figure 4.46). It should be also noted that, when the bottomhole pressure approaches P_c, the size of the initial straight-line section of the semilogarithmic dependence increases (Figure 4.44).

It is obvious that the observed phenomena are connected with the increase of the permeability and, correspondingly, medium piezo-conductivity in the region of pressures, where the slippage effect occurs. It can be clearly seen in Figure 4.47, which shows the calculation curve of pressure build-up, replotted in semi-logarithmic coordinates, and the dynamics of the porous medium permeability at $P/P_c = 1.3$. Furthermore, when the porous medium is stabilized, the intensity of pressure build-up decreases.

The obtained results agree well with the field data, presented in reference [36]. Figures 4.48 and 4.49 show the materials of Figure 2.6 from the mentioned work.

As seen from the figures, the pressure build-up curves (PBCs) for well X0 of the Bach Ho (White Tiger) oil field have the form of a polygonal line with two characteristic sections. Also, then the bottomhole pressure approximates P_c and the size of the initial straight-line section of the semi-logarithmic dependence increases (Figure 4.48). When approaching the critical region, the curves deviate to the left (i.e. toward smaller relaxation times) and, moreover, the pressure build-up time also decreases (Figure 4.49). Figure 4.50 shows the calculation and field PBCs based on the data of reference [36], re-plotted in semi-logarithmic coordinates. It can be seen from the figure that the proposed model qualitatively and quantitatively describes the field results.

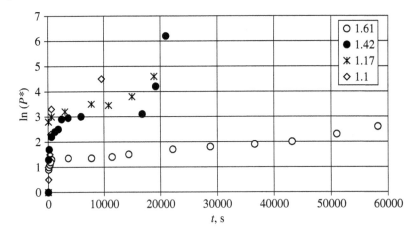

Figure 4.48 Pressure build-up curves for well X0 of the Bach Ho (White Tiger) oil field [36], replotted in semi-logarithmic coordinates at different P/P_c.

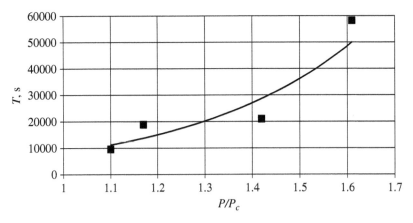

Figure 4.49 Dependence of pressure build-up time on P/P_c for well X0 of the Bach Ho (White Tiger) oil field [36].

In order to compare the results of the theoretical and experimental studies, the straight-line flow was considered. From the equations of the straight-line flow, and with the account of the initial and boundary conditions of Equations (4.72) and (4.73), as well as when applying the Kirchhoff transformation (4.75) and averaging method (4.76) for pressure distribution and build-up at the porous medium outlet at the pressure much above the saturation pressure, i.e. when there is no fluid slippage ($P > P_s$), the following equations were obtained:

$$P = \frac{x^2 - l^2}{2\chi_0} \frac{3Q_0}{2\beta^* f l} \exp\left(-\frac{3\chi_0}{l^2} t\right) \cdot \frac{3\chi_0}{l^2} \cdot t + P_0 + \frac{Q_0 \eta}{k_0 f} \exp\left(-\frac{3\chi_0}{l^2} t\right)(x - l) \quad (4.89)$$

$$P_e = P_0 - \frac{9Q_0}{4\beta^* f l} t \exp\left(-\frac{3\chi_0}{l^2} t\right) - \frac{Q_0 \eta l}{k_0 f} \exp\left(-\frac{3\chi_0}{l^2} t\right) \quad (4.90)$$

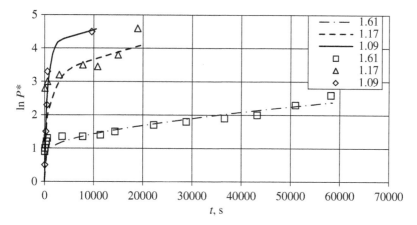

Figure 4.50 Calculated (lines) and original (points) pressure build-up curves for well X0 of the Bach Ho (White Tiger) oil field [36], replotted in semi-logarithmic coordinates at different P/P_c.

or

$$P_e = P_0 - \Delta P \exp\left(-\frac{3\chi_0}{l^2}t\right)\left(\frac{9\chi_0}{4l^2}t - \frac{1}{2}\right)$$

Pressure build-up in the presence of fluid slippage can be determined from the following equations:

at $P_c \leq P \leq P_m$

$$P_e = P_0 - \frac{R}{4a}\left[1 + \frac{4a}{R}(P_0 - P_c)\right] +$$

$$\sqrt{\left(\frac{R}{4a}\right)^2\left[1 + \frac{4a}{R}(P_0 - P_c)\right]^2 - \Delta P \exp\left(-\frac{3\chi_0}{l^2}t\right)\left[\frac{R}{2a}\left(\frac{9\chi_0}{4l^2}t + 1\right) - (P_0 + P_{e0} - 2P_c)\right]}$$

at $P_m \leq P \leq P_s$

$$P_e = P_0 + \frac{R}{a}\left[1 + \frac{a}{R}(P_s - P_0)\right] -$$

$$\sqrt{\left(\frac{R}{a}\right)^2\left[1 + \frac{a}{R}(P_s - P_0)\right]^2 + \Delta P \exp\left(-\frac{3\chi_0}{l^2}t\right)\left[\frac{2R}{a}\left(\frac{9\chi_0}{4l^2}t - \frac{1}{2}\right) - (P_0 + P_{e0} - 2P_s)\right]}$$

where P_{e0} is the initial pressure at the porous medium outlet and $\Delta P = P_0 - P_{e0}$.

Figures 4.51–4.54 show the results of the numerical calculation on the basis of the stated formulas with the following values of the parameters: $P_0 = 5.7$–10.2 MPa; $\eta = 1$ mPa s; $F = 8\times10^{-4}$ m^2; $\chi_0 = 0.05$ m^2/s; $k_0 = 0.15\times10^{-12}$ m^2; $a = 5\times10^{-11}$ m/Pa; $R = 2.45\times10^{-6}$ m; $\beta^* = 2.94\times10^{-9}$ 1/Pa; $l = 1.1$ m; $P_s = 7.8$ MPa; $P_c = 3$ MPa; $P_{e0} = 3.0$–7.5 MPa.

It can be seen from Figures 4.51–4.53 that the same consistencies are true for both the linear and the radial cases. At the same time, the obtained solution does not completely correspond to the experimental results, shown in Section 2.2. Due to the fact that the diffusion dissolution of the subcritical gas bubbles was not taken into account, the build-up

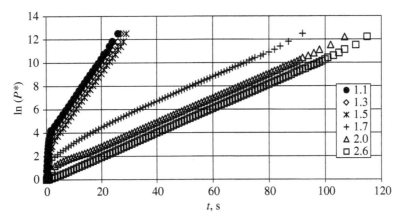

Figure 4.51 Calculated pressure build-up curves, replotted in semi-logarithmic coordinates at different P/P_c.

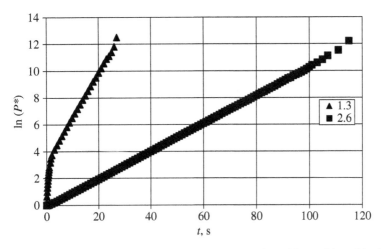

Figure 4.52 Calculated pressure build-up curves, replotted in semi-logarithmic coordinates at different P/P_c.

time decreases when approaching the critical region, as opposed to experimental studies. Differences between the solution for the radial case (where the diffusion gas dissolution was not taken into account either) and the field studies can be explained by the fact that the formation pressure does not build up completely during field studies. For its complete build-up, there is a need to shut in all the wells within the studied area, which is close to impossible to implement. The solution, obtained for the radial case here, can be used for interpretation of the field PBCs.

The process of diffusion gas dissolution was considered in reference [37]. It was shown that, in the presence of diffusion dissolution in the pressure build-up process, the corresponding pressure build-up curve sections in coordinates $\left(\ln\left(\sqrt{t}\,\partial P/\partial t\right),\ 1/t\right)$ are linear.

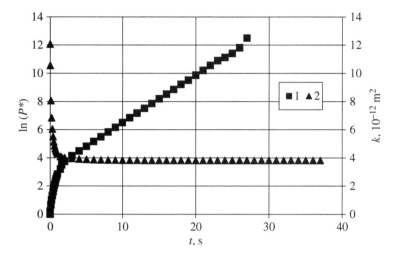

Figure 4.53 Calculated curve of pressure build-up, replotted in semi-logarithmic coordinates (1) and dynamics of porous medium permeability (2) at $P/P_c = 1.3$.

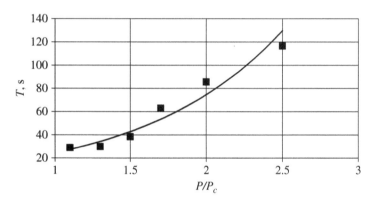

Figure 4.54 Dependence of pressure build-up time on P/P_c.

Figure 4.55 shows replotted experimental data on the pressure build-up when $\sqrt{t}\partial P/\partial t = P^0$. It can be seen from the figure that the final section becomes linear for the levels of initial pressure that are close to the critical region ($P = 1.3P_c$), but this does not happen for the levels $P \geq 2P_c$.

4.3.3 Viscosity Anomaly Near to the Phase Transition Point

Analysis of the typical viscosity dependence on pressure shows that the viscosity is reduced significantly when pressure decreases to the bubble point (saturation) pressure. Indeed, according to the results of studies, the viscosity of the live oil can be 20% lower at the pressure near the saturation pressure than at the reservoir pressure. This section describes studies of the gas saturated oil thermophysical properties and considers the influence of subcritical nucleation on the phase behavior and viscosity of the formation (e.g. live) oil.

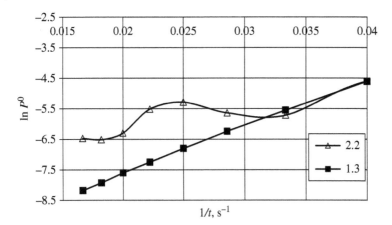

Figure 4.55 Experimental pressure build-up curves at different P/P_c, replotted in "diffusive" coordinates.

4.3.3.1 Experimental Procedures

The studies were conducted on live oil samples that were taken from well No. 6952 ("Uzen" field). Live oil sampling was provided according to Sampling Petroleum Reservoir Fluids, the American Petroleum Institute (API) Recommended Practice API RP 44. Table 4.1 shows the properties of the live oil.

The sulfonate micellar concentrate "Karpatol-UM2K-Nurol" was used. The concentrate is a mixture of ammonium or sodium sulfate, unsulfonated hydrocarbons, and oxygenated organic compounds. It is oil- and water-soluble. The reagent is produced by Karpatol LLC (Ukraine). Table 4.2 shows the main properties of the surfactant "Karpatol-UM2K-Nurol."

A study of the live oil phase behavior with and without surfactant additives was carried out using the Model 3000 GL Chandler Engineering PVT Phase Behavior System (see Figure 4.56) by the flash liberation test. Unlike typical studies [38–42] at a pressure near the saturation pressure, the measurements were conducted at the pressure increment of 0.08–0.25 MPa. The studies were conducted at the reservoir temperature of the "Uzen" field (66 °C).

Table 4.1 Properties of the oil.

Depth of sample collection (m)	500
Reservoir pressure (MPa)	12.42
Reservoir temperature (°C)	66
GOR (m^3/m^3)	23.95
Formation volume factor	1.098
Shrinkage (%)	8.92
Compressibility coefficient (1/MPa)	10.78×10^{-4}
Density (kg/m^3)	800.5
Density of separated oil at 20 °C (kg/m^3)	845.2

Table 4.2 Main parameters of the surfactant "Karpatol-UM2K-Nurol."

External view	Liquid, from light to dark brown color
Surfactants (wt.%)	≥ 29.0
Ammonium or sodium sulfate (wt.%)	≤10.0
Water (wt.%)	≤60.0
Unsulfonated hydrocarbons (wt.%)	≤30.0
pH	8–10
Density at 20 °C (g/cm³)	Not more than 1.15
Freezing temperature (°C)	Not higher than −10

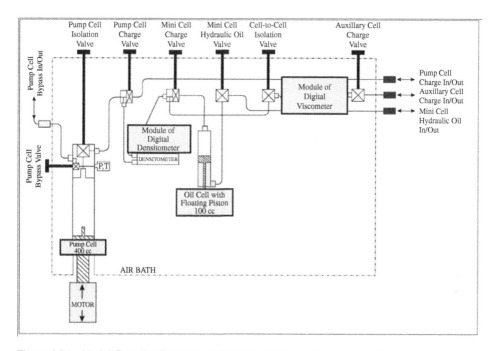

Figure 4.56 Model Chandler Engineering 3000 GL PVT Phase Bbehavior System scheme.

4.3.3.2 Measurement of Live Oil Viscosity

The viscosity of the live oil with and without surfactant additives at different pressures was measured by a Model 1602 Chandler Engineering Falling Ball Viscometer.

4.3.3.3 Phase Behavior of Live Oil and Viscosity Anomaly

Figure 4.57 and Table 4.3 show the experimental live oil viscosity dependence on pressure relative to the saturation pressure P/P_b (P_b – saturation pressure). It can be seen from the

4 Nanogas Emulsions in Oil Field Development

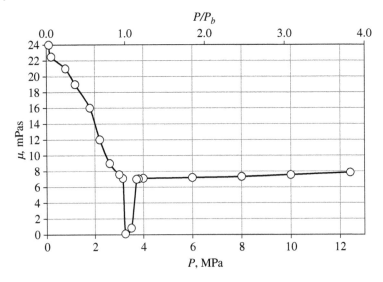

Figure 4.57 Live oil viscosity dependence on pressure and pressure level.

Table 4.3 Live oil viscosity dependence on pressure.

Pressure (MPa)	P/P_b	Viscosity (mPa s)
12.42	3.82	7.840
10.00	3.08	7.550
8.00	2.46	7.320
6.00	1.85	7.200
4.00	1.23	7.120
3.80	1.17	7.050
3.72	1.14	7.000
3.50	1.08	0.840
3.25	1.00	0.110
3.15	0.97	7.100
3.00	0.92	7.600
2.60	0.80	9.000
2.20	0.68	12.000
1.80	0.55	16.000
1.20	0.37	19.000
0.80	0.25	21.000
0.20	0.06	22.500
0.10	0.03	24.000

figures and the table that there is a significant viscosity anomaly in the pressure range $P/P_b = 1$–1.14. The viscosity in this range drops approximately 70 times in comparison to the viscosity at the reservoir pressure (12.42 MPa) and approaches gas viscosity. Figure 4.58 shows the dependence of the viscosity on the pressure without taking into account the two points in the range $P/P_b = 1$–1.14. It can be seen from the figure that the dependence of the viscosity on the pressure is typical without accounting for the anomalous region [43–47]. There is a decrease in the viscosity of 11% when the reservoir pressure decrement is up to 3.72 MPa. The pressure at the point of the minimum viscosity saturation pressure [38–46] was 3.25 MPa, whereas the saturation pressure is 3.72 MPa for a typical dependence (Figure 4.58), i.e. it is 15% higher than the actual saturation pressure.

The viscosity anomaly was observed on live oil samples from different fields and wells (Table 4.4).

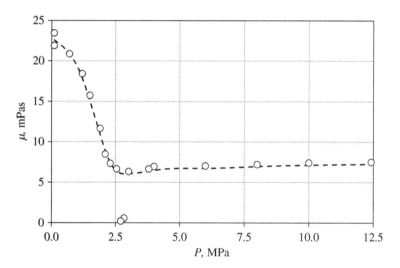

Figure 4.58 Typical live oil viscosity dependence on pressure.

Table 4.4 List of studied live oil samples.

Fields, well number	Date
Uzen, 4182	July, 2014
North Akkar, 4	October, 2014
Jetibay, 4482	October, 2014
Uzen, 6952	June, 2015

4.3.3.4 Surfactant Impact on Phase Behavior of Live Oil and Viscosity Anomaly

The water- and oil-soluble surfactant "Karpatol-UM2K-Nurol" with different concentrations was administered to the live oil under the reservoir pressure. The flash liberation test was used to determine the saturation pressure. Figure 4.59 shows the pressure–volume plot of the live oil without any additives and at the surfactant concentration of 23.3 wt.%. Figure 4.60 shows the surfactant concentration impact on the saturation pressure. It can be seen from the figure that the surfactant additive decreases the saturation pressure by almost 36%.

Figure 4.61 and Tables 4.3 and 4.5 show the experimental dependencies of the viscosity on the pressure without any additives and at the 5 wt.% surfactant concentration. It can be seen from the figures and the tables that there is a significant viscosity anomaly in the range of the pressure level $P/P_b = 1$–1.11. The viscosity in this range drops more than 37 times in

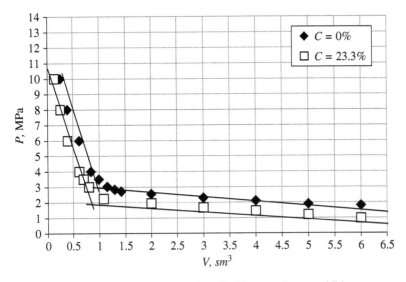

Figure 4.59 Pressure–volume plot with and without surfactant additives.

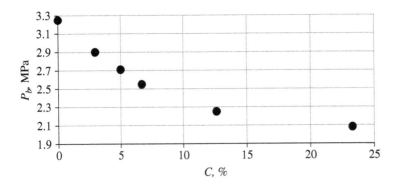

Figure 4.60 Dependence of saturation pressure on surfactant concentration.

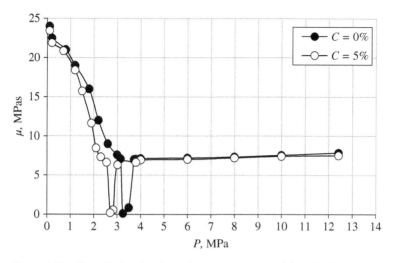

Figure 4.61 Live oil viscosity dependence on pressure with and without surfactant additives.

Table 4.5 Live oil viscosity dependence on pressure at the surfactant concentration of 5 wt.%.

Pressure (MPa)	P/P_b	Viscosity (mPa s)
12.42	4.58	7.510
10	3.69	7.420
8	2.95	7.210
6	2.21	7.030
4	1.48	6.950
3.8	1.40	6.650
3.02	1.11	6.320
2.84	1.05	0.560
2.71	1.00	0.200
2.55	0.94	6.650
2.3	0.85	7.330
2.1	0.77	8.470
1.9	0.70	11.630
1.5	0.55	15.740
1.2	0.44	18.410
0.7	0.26	20.850
0.2	0.04	21.870
0.1	0.04	23.450

comparison to the viscosity at the reservoir pressure. The pressure at the point of minimum viscosity–saturation pressure was 2.71 MPa, which is 17% lower than the one without any surfactant additives. At the same time, the saturation pressure is 3.02 MPa for a typical dependence (see Table 4.5) and is almost 12% higher than the actual saturation pressure.

Analysis of published data indicates that typical measurements of the saturation pressure are conducted with too large increments [38–42] and therefore the viscosity anomaly at the pressure near the bubble point pressure has not been detected yet. Smaller increments allowed the current study to observe the viscosity abnormality. According to references [3] and [4], the obtained results can be explained by the subcritical formation of the gas bubbles. Apart from the other phenomena, this process is also observed by the typical dependence of the gas saturated oil density on the pressure, which is also characterized by a reduction in the density with a simultaneous decrease from the reservoir to the saturation pressure [46]. It has been shown [7, 48] that the stabilization of the subcritical bubbles takes place due to the surfactant discharge on their surface. Some amounts of surfactants are always present in real systems, even without special treatment. According to reference [1], the surfactant presence in fluids does not remove the question of stability of subcritical bubbles since the surfactant additive can only reduce their dissolution rate, but cannot prevent it completely. Stabilization may occur due to electrical charges on the bubble surface [1, 49]. Furthermore, it is important to take into account the combined action of the interfacial tension and electrical charge.

4.3.3.5 Mechanism of Viscosity Anomaly

The fluid viscosity at high pressures and temperatures is generally measured in viscometers, which are based on determination of the length of the ball downfall into the fluid. The subcritical gas bubbles form at the pressure near the saturation pressure on the solid surface [4] of the viscometer ball and the slippage effect occurs [4]. Furthermore, the ball fall velocity in the oil increases and the fall time decreases.

Let us consider the ball downfall in the viscous fluid in order to prove the suggested mechanism. After neglecting the quadratic inertia terms and mass forces and taking the motion of the fluid particles to be axially symmetric for the differential equations of the viscous incompressible liquid motion in spherical coordinates, the following expression is obtained [10]:

$$\begin{cases} \dfrac{1}{R^2 \sin\theta} \dfrac{\partial^2 \psi}{\partial t\, \partial\theta} = \dfrac{1}{\rho} \dfrac{\partial P}{\partial R} + \dfrac{\nu}{R^2 \sin\theta} \dfrac{\partial D\psi}{\partial \theta} \\ \dfrac{1}{\sin\theta} \dfrac{\partial^2 \psi}{\partial t\, \partial R} = -\dfrac{1}{\rho} \dfrac{\partial P}{\partial \theta} + \dfrac{\nu}{\sin\theta} \dfrac{\partial D\psi}{\partial R} \end{cases} \quad (4.91)$$

where D is the Stokes operator

$$D = \frac{\partial^2}{\partial R^2} + \frac{\sin\theta}{R^2} \frac{\partial}{\partial \theta}\left(\frac{1}{\sin\theta} \frac{\partial}{\partial \theta}\right)$$

ψ is the stream function; R, θ are spherical coordinates; ρ is the liquid density; ν is the kinematic viscosity of liquid; and P is the pressure.

Radial (V_R) and tangential (V_θ) components of velocity of the liquid can be determined according to the following equations [50]:

$$V_R = -\frac{1}{R^2 \sin\theta} \frac{\partial \psi}{\partial \theta}$$

$$V_\theta = \frac{1}{R \sin\theta} \frac{\partial \psi}{\partial R} \quad (4.92)$$

When the pressure is removed from equation (4.91), the following differential equation for the stream function will be obtained:

$$\frac{\partial D\psi}{\partial t} = \nu DD\psi$$

Given that there is no slippage, the boundary conditions are in the form

$$V_R|_{R=a} = V_0 \cos\theta \quad (4.93)$$

$$V_\theta|_{R=a} = -V_0 \sin\theta \quad (4.94)$$

where V_0 is the ball velocity.

Based on the boundary conditions (4.93) and (4.94) and according to reference [50], the stream function is determined in the form

$$\psi = \sin^2\theta \, F(R; t) \quad (4.95)$$

where F is the function depending on the coordinate R and time t.

When the ball in the liquid falls with slippage, the boundary condition takes the form (see Figure 4.62)

$$b\frac{\partial V_\theta}{\partial R}\bigg|_{R=a-b} = -V_\theta|_{R=a-b} \quad (4.96)$$

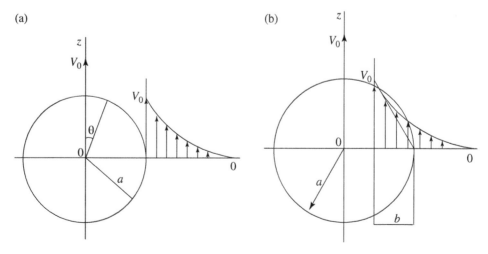

Figure 4.62 Diagram of velocity distribution during ball downfall: (a) without slippage; (b) with slippage.

where b is the slippage coefficient (Figure 4.62b). Furthermore, the condition without slippage is met not on the actual ball surface but on the surface equivalent radius $a_0 = a - b$ (Figure 4.62b).

The tangential component of the liquid velocity is determined from expression [10]

$$V_\theta = \frac{1}{R \sin \theta} \frac{\partial F}{\partial R} \tag{4.97}$$

In reference [50], the differential equations (4.91) were integrated, taking into account the boundary conditions (4.93) and (4.94) and expressions (4.92) and (4.95) at high values of time t and the velocity V_θ was then determined:

$$V_\theta = -\frac{a V_0 \sin \theta}{4R} \left(3 + \frac{a^2}{R^2}\right) \tag{4.98}$$

When inserting expression (4.98) into condition (4.96), the following expression is obtained:

$$3b^2 - 10\,a\,b + 7a^2 = 0 \tag{4.99}$$

From Equation (4.99) for the limit case, when the falling ball is fully covered with the boundary layer of gas and actually there is not any liquid resistance, $b = a$.

Then, for the resultant force acting on the ball that is moving linearly and homogeneously,

$$P_z = -6 \pi \mu\, a_0\, V_0 \left(1 + a_0 \sqrt{\frac{1}{\nu \pi t}}\right) \tag{4.100}$$

where μ is the dynamic viscosity, ν is the kinematic viscosity, and t is time. The second term in expression (4.100) can be neglected at the quite high t, which then proposes the Stokes formula.

The velocity of the ball downfall in the liquid at the homogeneous motion is calculated from the equation

$$mg - P_z - \frac{4}{3} \pi a_0^3 \rho_l g = 0 \tag{4.101}$$

From Equation (4.91) it is possible to obtain

$$V_0 = \frac{2 a_0^2 (\rho - \rho_l) g}{9 \mu}$$

where ρ is the density of the ball material and ρ_l is the liquid density.

Taking into account that $V_0 = H/t$, where H is the height of the homogeneous ball downfall, let us determine the fluid viscosity at the slippage

$$\mu = \frac{2 a_0^2 (\rho - \rho_l) g t}{9 H}$$

For the ratio of viscosities with slippage and without it, the following is obtained:

$$\frac{\mu_0}{\mu} = \frac{a_0^2}{a^2} = \left(1 - \frac{b}{a}\right)^2 \tag{4.102}$$

In the limit case $\mu_0 = 0$, but in practice the ball surface is only partially covered with nanobubbles as the occupancy of the surface with the nanobubbles can be only 20% according to the data of reference [18]. According to the data of the experiment shown above (see Figure 4.57), the slippage coefficient $b = 0.88a$ and the calculation based on formula (4.102) gives $\mu_0 = 0.014$, i.e. the viscosity at the saturation pressure decreases approximately 70 times in comparison with the viscosity at the reservoir pressure.

Based on the obtained results, it is possible to suggest the following mechanism of the viscosity anomaly at the pressure near the saturation pressure. When the pressure in the range $P/P_b = 3.82$–1.14 decreases, the boundary layer, which is partially made of stable subcritical gas bubbles, forms on the ball surface. This results in the slippage effect and leads to increment in the fall velocity. The result is that the apparent viscosity of the liquid in this range decreases by 11%. Further reduction of the pressure in the range $P/P_b = 1.14$–1 leads to an increase in the degree of the bubble cover on the ball surface, which strengthens the slippage effect [31]. The observed effect is a significant, seventy-fold, decrease in the apparent viscosity.

4.3.3.6 Mechanism of Surfactant Influence on Phase Behavior of Live Oil and Viscosity Anomaly

Equation (3.5) can be transformed to the following form:

$$\ln(P_b) = \ln(P) + \frac{2v_g \sigma_0}{kTr} - \frac{v_g(ze)^2}{16\pi^2 \varepsilon \varepsilon_0 r^4 kT}$$

The equation shows that the reduction of the surface tension results in the pressure decrease at other equal conditions. The experiments were carried out in the KRUSS DSA30 instrument with the view to determine the oil surface tension with a surfactant addition (Figure 4.63). It can be seen that the interfacial tension decreases when the surfactant concentration increases and at 23.3 wt.% it reaches the value that is practically 36% lower than without any surfactant additives.

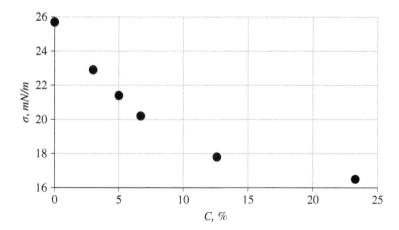

Figure 4.63 Dependence of oil surface tension on surfactant concentration.

It should be noted that the decrease in the viscosity in the anomaly zone is lower with the surfactant additives than without them. The viscosity in the anomaly zone decreases 70 times, but decreases only 37 times at 5 wt.% surfactant concentration.

It is known that nucleation of the new phase generally forms on the existing surfaces [51]. It is the surface of a steel ball in the viscometer in the considered case.

Indeed, the work of the heterogeneous formation of the nuclei can be determined from the ratio [17]

$$\frac{W_h}{W} = f(\theta) \qquad (4.103)$$

where W is the work of homogeneous formation of the gas phase bubble and $f(\theta)$ is a function of the contact angle of wetting of the solid surface by the liquid:

$$f(\theta) = \frac{1}{4}(1 + \cos\theta)^2(2 - \cos\theta) \qquad (4.104)$$

When inserting Equation (4.104) into Equation (4.103) and after some transformations, the following equation can be obtained:

$$\frac{W_h}{W} = \frac{1}{2} + \frac{3}{4}\cos\theta - \frac{1}{4}\cos^3\theta$$

It can be seen from this equation that the work of the heterogeneous process is always lower than the work of the homogeneous process at $0° < \theta \leq 180°$. Therefore, when adding the anionic surfactant, the wettability improves and less nuclei form on the ball surface, which leads to the decrease in the viscosity reduction. According to the data of the experiment shown above (see Figure 4.61), the slippage coefficient $b = 0.835a$ and the calculation based on formula (3.43) gives $\mu_0 = 0.027$; i.e. the viscosity in the anomaly zone at the saturation pressure decreases approximately 37 times as compared to the viscosity at the reservoir pressure.

4.3.3.7 Concluding Remarks

1) Based on the experimental studies, the viscosity anomaly was discovered in the range of the pressure $P/P_b = 1$–1.14. The viscosity near the bubble point of the live oil decreases almost 70 times when compared to the viscosity at the reservoir pressure.
2) The influence of the surfactant on the phase behavior of the live oil and viscosity anomaly were studied experimentally. It was found that the saturation pressure decreases significantly (up to 36%) when surfactant is added. Furthermore, the viscosity of the live oil at the surfactant concentration of 5 wt.% decreases almost 37 times when compared to the viscosity at the reservoir pressure and the anomality pressure.
3) The mechanism of the viscosity anomaly was suggested based on formation of the stable subcritical gas bubbles and associated slippage effect.
4) The model for determining the oil viscosity taking into account the slippage effect was suggested.

5

Nanoaerosoles in Gas Condensate Field Development

A gas condensate fluid is a hydrocarbon aerosol dispersion system in which the gas and liquid are the dispersion and dispersed phases respectively. This mixture behavior has a significant impact on flow in hydrocarbon aerosol porous media. Indeed, liquid secretion occurs in a porous medium when the pressure drops below the dewpoint pressure, which leads to a sharp decrease in the relative gas permeability, ultimately restricting the gas flow.

Thus, the study of hydrocarbon aerosol flows in porous media is of significant interest, especially when subcritical gas-condensate fluids are involved. This chapter presents a mechanism for subcritical condensate nucleus stabilization via experimental and theoretical studies of steady- and unsteady-state gas condensate fluid flows above the dewpoint pressure in porous media.

Presented results have a significant value from the theoretical and practical viewpoints and can find wide applications in gas fields development.

A recombined gas condensate mixture consisting of natural gas with a density of 0.745 kg/m^3 and n-hexane was studied (dewpoint pressure: 17.5 MPa, temperature: 333 K, gas/condensate ratio: 4800 Nm^3/m^3). Table 5.1 shows the natural gas composition.

An organosilicon amide of perfluorocarboxylic acid (Biohim, LLC) was used as an oleophobic agent.

5.1 Study of the Gas Condensate Flow in a Porous Medium

Experiments were performed using the setup shown in Figure 5.1. The setup incorporated the following elements: a sand pack, pressure gauges, a floating piston accumulator, a syringe pump, a temperature controller, a gasometer, a separator, a backpressure regulator, a pressure transducer, and a datalogger.

The experiments were conducted in the following sequence:

- A high-pressure sand-pack column with a length of 1.1 m and an inner diameter of 0.032 m was filled with variable fractions of quartz sand (average fraction diameter of 1.4×10^{-5} m).
- The pore volume and absolute permeability were determined in accordance with the standard procedure (the absolute permeability was 0.02 µm^2).

Nanocolloids for Petroleum Engineering: Fundamentals and Practices, First Edition.
Baghir A. Suleimanov, Elchin F. Veliyev, and Vladimir Vishnyakov.
© 2022 John Wiley & Sons Ltd. Published 2022 by John Wiley & Sons Ltd.

Table 5.1 Natural gas composition.

Components	mol. (%)
Carbon dioxide	0.32
Methane	95.57
Ethane	2.75
Propane	0.75
Iso- butane	0.13
Normal butane	0.22
Iso-pentane	0.09
Normal pentane	0.08
Normal hexane	0.07
Heptane+	0.02

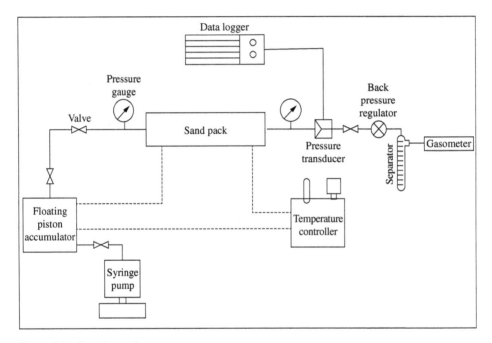

Figure 5.1 Experimental setup.

- The experimental setup was vacuum-treated under constant thermostatic control ($T = 333$ K).
- A recombined gas condensate mixture was prepared in the floating piston accumulator.
- The sand pack was saturated with natural gas, which was then displaced by the gas condensate mixture at 37.0 MPa.

- Displacement continued until the gas/condensate ratio of the mixture reached equilibrium.
- The sand pack outlet was shut off for 48 hours.
- Various pressures were created at the sand pack inlet and outlet (P_i and P_o respectively). The gas condensate was filtered under a constant pressure drop of 0.8 MPa until a constant gas flow rate was established. The dependence of the gas flow rate on the average pressure in the sand pack $P = (P_i + P_o)/2$ was determined.
- The sand pack outlet was shut off and the pressure build-up curve was measured.

After the steady-state and unsteady-state studies were conducted, the pressure was reduced to the next level and similar measurements were performed. These experiments were continued until the pressure reached 20.8 MPa. Thus, the studies were performed within the $P/P_c = 1.2 - 2$ pressure range (P_c is the dewpoint pressure). The results are expressed in the form of the dependence of the gas flow rate on the average pressure and pressure build-up curves.

The dependence of the compressibility coefficient on the pressure in the porous medium was determined in the next series of experiments. The experiments were conducted in the following sequence:

- The sand pack was saturated with natural gas, which was then displaced by the gas condensate mixture at 37.0 MPa.
- Displacement continued until the mixture gas/condensate ratio reached equilibrium.
- The sand pack outlet was shut off for 48 hours.
- The sand pack outlet was opened to release the gas condensate mixture until the target pressure was reached. The volume of the gas condensate mixture, which was taken from and remained in the porous medium, was measured.
- The pressure in the sand pack was decreased to the next level and similar measurements were performed.
- The compressibility coefficient was determined according to the formula

$$\beta = -\frac{\Delta V}{V \Delta P} \tag{5.1}$$

where V is the volume of the sample in the accumulator at pressure P and ΔV is the change in the sample volume due to the pressure change ΔP at constant temperature. The results of these experiments were expressed as the dependence of the compressibility coefficient on the pressure. The wetting contact angle was measured using a KRUSS DSA30 instrument.

Figure 5.2 shows the dependence of the gas flow rate (and relative gas flow rate $Q_0 = Q/Q_p = 32\text{ MPa}$, shown by the dashed line) on the pressure. When the relative pressure is at 1.5 (27.2 MPa), the gas flow rate is almost 30% higher than at 20.8 MPa and 25% higher than at 32 MPa. The fluid flow rate starts to increase at 1.74 times (30.4 MPa) the dewpoint pressure. Furthermore, the dependence of the gas flow rate on the pressure is nonmonotonic and increased flow rates are achieved within the $P/P_c = 1.4-1.7$ (24.5–30 MPa) range.

Figure 5.3 shows pressure build-up curves taken at 32, 27.2, and 16.5 MPa (below the dewpoint pressure). The pressure decrease leads to considerable increases in the pressure build-up time and hydraulic diffusivity (see Table 5.2).

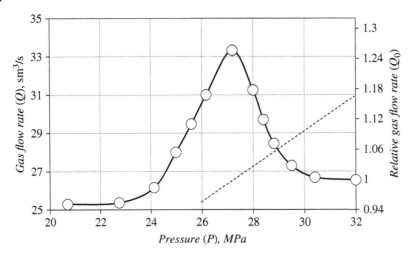

Figure 5.2 Dependence of the gas flow rate on the pressure.

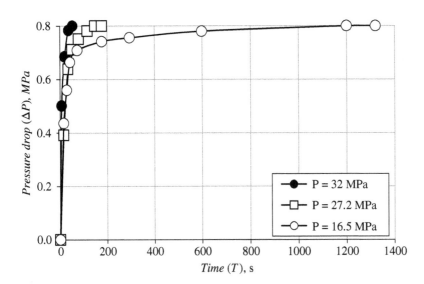

Figure 5.3 Pressure build-up curves.

Table 5.2 Hydraulic diffusivities at various pressures.

Pressure (MPa)	The hydraulic diffusivity (m² s⁻¹)
32 (32.4–31.6)	0.032
27.2 (27.6–26.8)	0.015
16 (16.4–15.6)	0.006

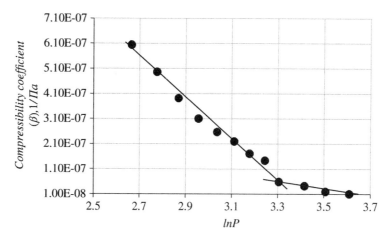

Figure 5.4 Dependence of the compressibility coefficient on the pressure in semi-logarithmic coordinates.

Figure 5.4 shows the dependence of the compressibility coefficient on the pressure, replotted using semi-logarithmic coordinates. The compressibility coefficient increases significantly when the pressure decreases. Moreover, two stages characterized by different levels of compressibility coefficient growth can be distinguished using semilogarithmic coordinates. The first stage begins at 37 MPa and ends at 27.2 MPa, where the maximum gas flow rate is achieved. The second stage includes the dewpoint pressure and is characterized by a larger compressibility coefficient increase. The resulting dependence can be described by the sum of two exponents according to the formula

$$\beta = A_1 \exp(-B_1 P) + A_2 \exp(-B_2 P); A_1 = 1.209 \times 10^{-6}; B_1 = 0.127; \\ A_2 = 5.367 \times 10^{-6}; B_2 = 0.176 \tag{5.2}$$

5.2 Mechanism of the Gas Condensate Mixture Flow

Figure 5.5 shows the phase diagram (isotherm) of the gas condensate mixture obtained using the reservoir model (see Figure 5.1) under the experimental conditions described above. Under the traditional approach, the system is a single phase (liquid-saturated gas) above the dewpoint and contains two phases (gas and liquid condensate) below it [52].

Under the traditional approach, gas flow in a porous medium takes place above the dewpoint pressure. Meanwhile, the volumetric flow rate can be determined using Darcy's law [53]

$$Q = \frac{kF}{2\eta P_{at}} \frac{\left(P_i^2 - P_o^2\right)}{l} \tag{5.3}$$

where k is the permeability, F is the cross-sectional area, l is the length of the porous medium, η is the gas viscosity, and P_{at} is the atmospheric pressure.

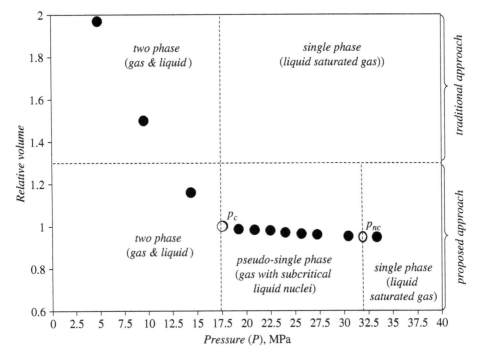

Figure 5.5 Gas condensate phase diagram (isotherm).

According to the Equation (5.3) (Figure 5.2, dashed line), the gas flow decreases with the average pressure. These results cannot be explained using the traditional approach. The following approach is proposed to explain the results based on the idea that nucleation theory is applicable to both single-component pure and multi-phase (gas emulsion, aerosol, etc.) systems [3, 54, 55]. The presence of three regions is assumed (Figure 5.5): the $p > p_{nc}$ single-phase region (liquid-saturated gas), wherein p_{nc} is the pressure when the subcritical condensation nucleus starts to release, and is identified experimentally in Figure 5.2, i.e. $p_{nc} = 32$ MPa; the $p_{nc} \geq p \geq p_c$ pseudo-single-phase region (gas with subcritical liquid nuclei); and the $p < p_c$ two-phase region (gas and liquid).

The presence of subcritical liquid nuclei has been proven by several studies, and has been demonstrated with water vapor [3]. The practical aspects of fluid subcritical nucleus formation and stabilization have not been studied. Hence, a mechanism was suggested for subcritical cluster stabilization, as well as a mechanism for gas condensate flow in porous media in the presence of subcritical liquid nuclei.

5.2.1 Rheology Mechanism of the Gas Condensate Mixture During Steady-State Flow

It is known that the nuclei of new phases (bubbles or liquid droplets) generally form on existing surfaces [51]. Furthermore, the work of a heterogeneous process is always less than the work of a homogeneous process. It was assumed that the subcritical liquid nuclei that form (adsorb) on the capillary surfaces are mobile, and hence that the boundary layer has a higher viscosity than the gas in the center of the flow. It is natural to assume that the volumetric content increases when the pressure decreases to the dewpoint pressure, owing to a decrease in the work of formation of the nuclei, thus increasing the condensate boundary layer thickness. It was not considered to be the Klinkenberg effect [56] since its influence on natural gas is reduced to nearly zero at 2 MPa.

5.2.1.1 Annular Flow Scheme in a Porous Medium Capillary

Let us consider a gas condensate flow in an ideal porous medium. The medium is represented as a bundle of cylindrical capillaries with the same diameter, through which annular flow occurs (see Figure 5.6a). Furthermore, the gas phase moves in the center of the flow and the fluid phase moves near the capillary wall. The steady-state velocity of the fluid flow in the capillary under the inertia less approximation is determined from the equation

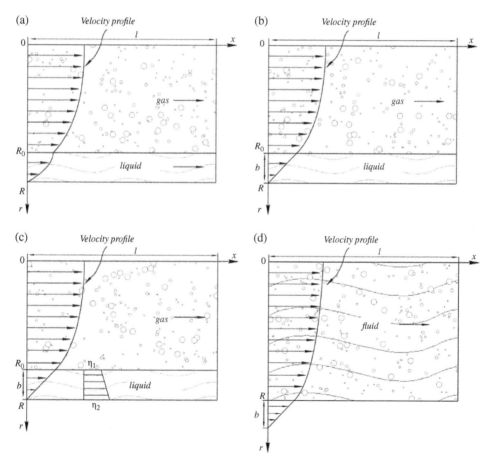

Figure 5.6 The annular flow models in a capillary: (a) without slip, (b) with slip, (c) with variable viscosity, and (d) interpretation of the Maxwell–Navier slip length for a real fluid (partial slip).

$$\frac{1}{r}\frac{d}{dr}\left(r\eta_i \frac{dv_i}{dr}\right) = -\frac{dp}{dx} \quad \left(i = \begin{cases} 1, & 0 \le r \le R_0 \\ 2, & R_0 \le r \le R \end{cases}\right) \quad (5.4)$$

where v_i is the distribution of velocities along the capillary radius r for the cross-section with the coordinate x, p is the pressure, and η_i is the dynamic viscosity.

Solving Equation (5.4) using the following boundary conditions:

$$v_2 = 0, \ r = R; \ \eta_2 \frac{dv_2}{dr} = \eta_1 \frac{dv_1}{dr}, \ v_1 = v_2, \ r = R_0; \ \frac{dv_1}{dr} = 0, \ r = 0$$

produces

$$v_1 = \frac{1}{4l\eta_1}\frac{dp}{dx}\left[R_0^2 - r^2 + \varepsilon\left(R^2 - R_0^2\right)\right] \quad (5.5)$$

where l is the capillary length.

The gas flow rate is determined from the expression

$$Q_{ns} = 2\pi \int_0^{R_0} rv_1 \, dr = \frac{\pi R_0^4}{8\eta_1}\frac{dp}{dx}\left[1 + 2\varepsilon\frac{R^2 - R_0^2}{R_0^2}\right] \quad (5.6)$$

and the relative gas flow rate is

$$Q_0 = \frac{Q_{ns}}{Q} = S^4\left[1 + 2\varepsilon\frac{1 - S^2}{S^2}\right] = (1 - \xi)^4\left[1 + 2\varepsilon\frac{\xi(2 - \xi)}{(1 - \xi)^2}\right] \quad (5.7)$$

where $\xi = \delta/R$, $\delta = R - R_0$, $S = R_0/R = 1 - \xi$, $\varepsilon = (\eta_1/\eta_2) < 1$, and Q is the Poiseuille flow rate of a gas with viscosity η_1.

The dependence of the volume concentration C of the liquid phase (or condensate saturation) on the pressure in the region $p_{nc} \ge p \ge p_c$ can be determined using the first approximation

$$C = C_0 \frac{p_{nc} - p}{p_{nc} - p_c} \quad (5.8)$$

where C_0 is the concentration of the condensate at $p = p_c$.

The volumetric concentration of the boundary layer (or condensate saturation) can be determined using the following equation:

$$C = \frac{\pi(R^2 - R_0^2)l}{\pi R^2 l}f = (1 - S^2)f \quad (5.9)$$

where f is a constant coefficient that characterizes the degree to which the condensate covers the inner capillary surface.

Then, by combining Equations (5.9) and (5.8), the following equation is obtained:

$$S = 1 - \zeta = \sqrt{1 - \frac{C_0}{f}\frac{p_{nc} - p}{p_{nc} - p_c}} \quad (5.10)$$

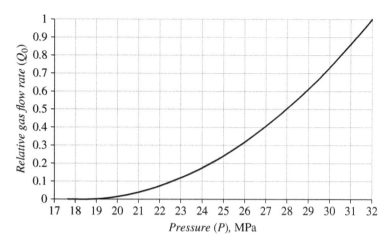

Figure 5.7 Calculated dependence of the relative gas flow rate on the annular flow pressure.

Inserting Equation (5.10) into Equation (5.7) produces

$$Q_0 = \left[1 - \frac{C_0}{f}\frac{p_{nc}-p}{p_{nc}-p_c}\right]^2 \left[1 + 2\varepsilon \frac{\frac{C_0}{f}\frac{p_{nc}-p}{p_{nc}-p_c}}{1 - \frac{C_0}{f}\frac{p_{nc}-p}{p_{nc}-p_c}}\right] \tag{5.11}$$

Figure 5.7 shows the dependence of the relative gas flow rate on the pressure, as calculated in accordance with Equation (5.11). It can be seen from the figure that the gas flow rate at any pressure level is lower than the Poiseuille flow rate. Thus, the annular flow scheme in the capillary does not describe the experimental results obtained.

5.2.1.2 Slippage Effect

Now consider an annular gas condensate flow in the capillary under an inertia-less approximation in the presence of slippage (see Figure 5.6b). Upon solving Equation (5.4) with the following boundary conditions:

$$v_1 = -b\frac{dv_1}{dr}, r = R_0; \quad \frac{dv_1}{dr} = 0, r = 0$$

then

$$v_1 = \frac{R_0^2}{4\eta_1}\frac{dp}{dx}\left[\left(\frac{r^2}{R_0^2} - 1\right) + \frac{2b}{R_0}\right]$$

where b is the slippage coefficient.

The gas flow rate with slippage is determined from the expression

$$Q_0 = 2\pi \int_0^{R_0} rv_1 \, dr = \frac{\pi R_0^4}{8\eta_1}\frac{dp}{dx}\left(1 + \frac{4b}{R_0}\right)$$

Considering the determination of the slippage coefficient [57] (see Figure 5.6d) and the fact that adhesion ($v_2 = 0$) occurs on the wall of the capillary ($r_n = R$), it was assumed that $b = \delta = R - R_0$. Finally, for the gas flow rate with slippage,

$$Q_0 = \frac{\pi R_0^4}{8\eta_1} \frac{dp}{dx} \left(1 + \frac{4(1-S)}{S}\right) \qquad (5.12)$$

The relative gas flow rate, which is the ratio of the gas flow rate with slippage (Equation (5.12)) to that without slippage (Equation (5.6)), is determined using the expression

$$Q_0 = \frac{4S - 3S^2}{S^2 + 2\varepsilon(1 - S^2)} = \frac{(1-\xi)(1+3\xi)}{(1-\xi)^2 + 2\varepsilon(2\xi - \xi^2)} \qquad (5.13)$$

Inserting Equation (5.10) into Equation (5.13) gives

$$Q_0 = \frac{4\sqrt{1 - \dfrac{C_0}{f} \dfrac{p_{nc} - p}{p_{nc} - p_c}} - 3\left[1 - \dfrac{C_0}{f} \dfrac{p_{nc} - p}{p_{nc} - p_c}\right]}{1 - \dfrac{C_0}{f} \dfrac{p_{nc} - p}{p_{nc} - p_c} + 2\varepsilon \left[1 - \sqrt{1 - \dfrac{C_0}{f} \dfrac{p_{nc} - p}{p_{nc} - p_c}}\right]^2}$$

Some studies [9, 10] have shown that even a smooth surface is only partially covered by subcritical liquid nuclei. Moreover, the number of nuclei formed on a surface with nanoscale roughness is lower. According to a previous study, the occupancy of a surface with such nuclei is limited to approximately 20%. It is obvious that lower coverage values are also possible in porous media. The approach proposed in reference [58] could be used to determine the fluid flow rate with intermittent coverage of the capillary surface with condensate nuclei. The cumulative flow velocity can be determined using the following equation:

$$v_c = f v_1 + (1 - f) v_0 \qquad (5.14)$$

where v_1 is the velocity, as determined using Equation (5.5), and v_0 is the Poiseuille velocity. From Equation (5.14),

$$v_c = \frac{1}{4\eta_1} \frac{dp}{dx} [f(R_0^2 - r^2 + \varepsilon(R^2 - R_0^2)) + (1-f)(R^2 - r^2)]$$

The cumulative gas flow rate without slip is determined using the expression

$$Q_c = 2\pi \int_0^{R_0} r v_c \, dr = \frac{\pi R_0^4}{8\eta_1} \frac{dp}{dx} \left[f\left(1 + 2\varepsilon \frac{1-S^2}{S^2}\right) + (1-f)\left(\frac{2-S^2}{S^2}\right) \right]$$

$$= \frac{\pi R_0^4}{8\eta_1} \frac{dp}{dx} \left[f\left(1 + 2\varepsilon \frac{2\xi - \xi^2}{(1-\xi)^2}\right) + (1-f)\left(\frac{1 + 2\xi - \xi^2}{(1-\xi)^2}\right) \right] \qquad (5.15)$$

The relative gas flow rate, which is the ratio of the gas flow rates calculated with (Equation (5.12)) and without (Equation (5.15)) slippage, is determined using the expression

5.2 Mechanism of the Gas Condensate Mixture Flow

$$Q_0 = \frac{(4-3S)/S}{f\left(1+2\varepsilon\frac{1-S^2}{S^2}\right)+(1-f)\left(\frac{2-S^2}{S^2}\right)} = \frac{(1+3\zeta)/(1-\zeta)}{f\left(1+2\varepsilon\frac{2\zeta-\zeta^2}{(1-\zeta)^2}\right)+(1-f)\left(\frac{1+2\zeta-\zeta^2}{(1-\zeta)^2}\right)}$$

(5.16)

Equation (5.16) becomes Equation (5.13) when $f = 1$. Upon inserting Equation (5.10) into Equation (5.16), the following equation is obtained:

$$Q_0 = \frac{4 - 3\left[\sqrt{1 - \dfrac{C_0}{f}\dfrac{p_{nc}-p}{p_{nc}-p_c}}\right]}{\sqrt{1 - \dfrac{C_0}{f}\dfrac{p_{nc}-p}{p_{nc}-p_c}}}{f\left\{1 + 2\varepsilon\dfrac{\dfrac{C_0}{f}\dfrac{p_{nc}-p}{p_{nc}-p_c}}{1 - \dfrac{C_0}{f}\dfrac{p_{nc}-p}{p_{nc}-p_c}}\right\} + (1-f)\dfrac{1 + \dfrac{C_0}{f}\dfrac{p_{nc}-p}{p_{nc}-p_c}}{1 - \dfrac{C_0}{f}\dfrac{p_{nc}-p}{p_{nc}-p_c}}}$$

The wall layer has a variable viscosity even in the one-phase system, e.g. water in fine pores [59, 60]. In the considered case, the nuclei form within the gas volume on the surface of the pore channel. Furthermore, it is natural to assume that the viscosity on the contact line between the condensate clusters, which cover the surfaces of the pore channel and gas, is almost equal to the gas viscosity and that the viscosity on the capillary wall equals the condensate viscosity. Hence, the following relationship for the viscosity distribution along the capillary radius to a first approximation could be analyzed (see Figure 5.6c):

$$\eta(r) = \eta_1 + \frac{\eta_2 - \eta_1}{R - R_0}(r - R_0) \tag{5.17}$$

Solving Equation (5.4) with the aid of Equation (5.17) produces

$$Q_1 = \frac{\pi R_0^4}{8\eta_1}\frac{dp}{dx}\left\langle \eta_1 f \left\{ \left[-\frac{1}{\eta_1} + 2\varepsilon\frac{2\left(\frac{1}{\varepsilon}-1\right)}{(\eta_2-\eta_1)S} - \eta_1\right] \ln\left(\frac{\eta_1(1-S)}{\eta_2-\eta_1} \atop S - S^2 + \frac{\eta_1(S-S^2)}{\eta_2-\eta_1}\right) + \frac{1}{\varepsilon\eta_1}\right] + \right.\right.$$
$$\left.\left.\frac{4(1-S)}{(\eta_2-\eta_1)S^2}\left[1 - S - \ln\left(\frac{\eta_1(1-S)}{\eta_2-\eta_1} \atop 1 - S + \frac{\eta_1(1-S)}{\eta_2-\eta_1}\right)\left(S - \frac{\eta_1(1-S)}{\eta_2-\eta_1}\right)\right]\right.\right.$$
$$\left.\left.+ \frac{2-S^2}{S^2}(1-f)\right\rangle$$

$$Q_0 = \dfrac{\dfrac{1}{\eta_1}\left(1+4\dfrac{1-S}{S}\right)}{f\left\{\left[-\dfrac{1}{\eta_1}+2\varepsilon\left|\dfrac{2\left(\dfrac{1}{\varepsilon}-1\right)}{\dfrac{(\eta_2-\eta_1)S}{1-S}-\eta_1}\right|\ln\dfrac{\dfrac{\eta_1(1-S)}{\eta_2-\eta_1}}{S-S^2+\dfrac{\eta_1(S-S^2)}{\eta_2-\eta_1}}+\dfrac{1}{\varepsilon\eta_1}\right] + \dfrac{4(1-S)}{(\eta_2-\eta_1)S^2}\left[1-S-\ln\dfrac{\dfrac{\eta_1(1-S)}{\eta_2-\eta_1}}{1-S+\dfrac{\eta_1(1-S)}{\eta_2-\eta_1}}\right]\left(S-\dfrac{\eta_1(1-S)}{\eta_2-\eta_1}\right)\right\} + \dfrac{2-S^2}{\eta_1 S^2}(1-f)}$$

(5.18)

$$Q_2 = \dfrac{\pi R^4}{\eta_2}\dfrac{dP}{dx}\dfrac{(1-S)}{(1-\varepsilon)} \left\{ \dfrac{1-3S^2+2S^3}{6} + \left[S^2\ln\dfrac{\dfrac{\eta_1(1-S)}{\eta_2-\eta_1}}{1-S+\dfrac{\eta_1(1-S)}{\eta_2-\eta_1}} + \dfrac{1-S^2}{2} + \right.\right.$$
$$\left. \left(S-\dfrac{\eta_1(1-S)}{\eta_2-\eta_1}\right)\left[1-S-\left(S-\dfrac{\eta_1(1-S)}{\eta_2-\eta_1}\right)\right]\ln\left(\dfrac{1-S+\dfrac{\eta_1(1-S)}{\eta_2-\eta_1}}{\dfrac{\eta_1(1-S)}{\eta_2-\eta_1}}\right)\right]$$

$$\left[\left(\dfrac{S}{2}+\dfrac{\eta_1(1-S)}{2(\eta_2-\eta_1)}\right) - \dfrac{S\varepsilon}{2}\left(\dfrac{1}{\varepsilon}-1\right)\dfrac{1}{1-\dfrac{\eta_1(1-S)}{S(\eta_2-\eta_1)}}\right]$$

$$\left. -\dfrac{S^2\varepsilon}{2}\left(\dfrac{1}{\varepsilon}-1\right)\dfrac{1}{1-\dfrac{\eta_1(1-S)}{S(\eta_2-\eta_1)}}\left(S\ln S + S\dfrac{(1-S^2)}{2}\right)\right\}$$

(5.19)

Numeric calculations were performed using Equation (5.18) with Equation (5.10) and the following parametric values: $C_0 = 0.6$; $f = 0.45$; $\varepsilon = 0.054$; $P_c = 17.5$ MPa; $\eta_1 = 1.2 \times 10^{-5}$ Pa s; $\eta_2 = 22 \times 10^{-5}$ Pa s

Figure 5.8 shows the results of the calculations. As can be seen from the figure, the calculation based on Equation (5.18) qualitatively describes the non-monotonic evolution of the gas flow rate and its maximum value. It is possible to suggest the following mechanism for the effects observed. The stable condensate nuclei, whose sizes range from 1 to 10 nm and which nucleates first on the wall layer and then effectively defines the wall, define the gas condensate flows in the subcritical region on the pore channel surfaces at a wetting contact angle $\theta > 0°$. This causes a slippage effect and the gas flow rate increases when the pressure decreases to the dewpoint pressure. At this point, the gas flow area, which is proportional to the gas flow rate, decreases because the condensate-saturated boundary layer thickens. The competition between these two effects produces a non-monotonic relationship between the fluid flow rate and the pressure. Furthermore, the slippage effect

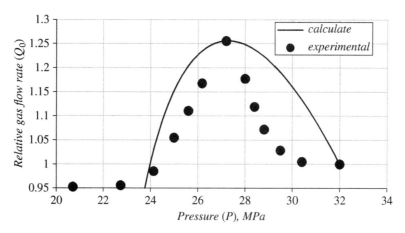

Figure 5.8 Calculated dependence of the relative gas flow rate on the pressure.

dominates before the maximum is reached and the subsequent flow rate decrease is connected to the decrease in the wetted cross-section of the pore channel, which is driven by increases in the boundary layer thickness.

It is helpful to determine the relative permeabilities in advance in order to analyze the real porous medium. Combining Equations (5.14) and (5.11) produces

$$k_1 = \left(1 - \frac{C}{f}\right)^2 \left(1 + 4\frac{1 - \sqrt{1 - \frac{C}{f}}}{\sqrt{1 - \frac{C}{f}}}\right) \tag{5.20}$$

Combining Equations (5.19) and (5.11) produces

$$k_2 = \frac{1 - \sqrt{1 - \frac{C}{f}}}{1 - \varepsilon} \left\{ \begin{array}{l} \dfrac{\dfrac{C}{f} + 2\left(1 - \dfrac{C}{f}\right)^{\frac{3}{2}}}{6} + \\[2ex] \left[\left(1 - \dfrac{C}{f}\right)\ln\dfrac{\eta_1\left(1 - \sqrt{1 - \dfrac{C}{f}}\right)}{\eta_2 - \eta_1} + \dfrac{C}{2f} + \right. \\[2ex] \left(\sqrt{1 - \dfrac{C}{f}} - \dfrac{\eta_1\left(1 - \sqrt{1 - \dfrac{C}{f}}\right)}{\eta_2 - \eta_1}\right)\left[\left(1 - \sqrt{1 - \dfrac{C}{f}}\right) - \left(\sqrt{1 - \dfrac{C}{f}} - \dfrac{\eta_1\left(1 - \sqrt{1 - \dfrac{C}{f}}\right)}{\eta_2 - \eta_1}\right)\right] \\[2ex] \left. \times \ln\dfrac{\left(1 - \sqrt{1 - \dfrac{C}{f}}\right) + \dfrac{\eta_1\left(1 - \sqrt{1 - \dfrac{C}{f}}\right)}{\eta_2 - \eta_1}}{\dfrac{\eta_1\left(1 - \sqrt{1 - \dfrac{C}{f}}\right)}{\eta_2 - \eta_1}} \right] \end{array} \right\}$$

$$\left[\frac{\sqrt{1-\frac{C}{f}}}{2} + \frac{\eta_1\left(1-\sqrt{1-\frac{C}{f}}\right)}{2(\eta_2-\eta_1)} - \frac{\sqrt{1-\frac{C}{f}}\varepsilon}{2}\left(\frac{1}{\varepsilon}-1\right)\frac{1}{1-\frac{\eta_1\left(1-\sqrt{1-\frac{C}{f}}\right)}{(\eta_2-\eta_1)\sqrt{1-\frac{C}{f}}}}\right.$$

$$\left. - \frac{\left(1-\frac{C}{f}\right)\varepsilon}{2}\left(\frac{1}{\varepsilon}-1\right)\frac{1}{1-\frac{\eta_1\left(1-\sqrt{1-\frac{C}{f}}\right)}{(\eta_2-\eta_1)\sqrt{1-\frac{C}{f}}}}\left(\sqrt{1-\frac{C}{f}}\ln\left(\sqrt{1-\frac{C}{f}}\right) + \sqrt{1-\frac{C}{f}\frac{C}{2f}}\right)\right]$$

(5.21)

Upon using the previously mentioned relative permeabilities in Darcy's generalized equation, it is possible to determine the main process parameters in the real porous medium under slippage with a variable-viscosity boundary layer [61].

5.2.2 Mechanism of Porous Medium Wettability Influence on the Steady-State Gas Condensate Flow

The nuclei of the new phase generally form on existing surfaces [51]. Indeed, the work of heterogeneous nucleus formation can be determined from the ratio [17]

$$\frac{W_h}{W} = f(\theta) \qquad (5.22)$$

where W is the work of the homogeneous formation of the fluid phase nuclei and $f(\theta)$ is the function that describes the wetting contact angle between the solid surface and the liquid:

$$f(\theta) = \frac{1}{4}(1-\cos\theta)^2(2+\cos\theta) \qquad (5.23)$$

Inserting Equation (5.23) into Equation (5.22) produces

$$\frac{W_h}{W} = \frac{1}{2} - \frac{3}{4}\cos\theta + \frac{1}{4}\cos^3\theta \qquad (5.24)$$

The work of the heterogeneous process is always smaller than the work of the homogeneous process at $0° < \theta < 180°$. Furthermore, when the porous medium (which consists of quartz sand in the considered case) is wetted by the condensate ($0° < \theta < 90°$), intense nucleation takes place on the porous channel surface, causing the slippage effect. When the porous medium is not wetted by the condensate ($90° < \theta < 180°$), little nucleation occurs on the porous channel surface and the porous medium becomes wet with the gas. As a result, the gas removes all of the condensate from the porous medium and no slippage effect is observed. A third set of experiments was performed to confirm this hypothesis.

The experiment was conducted in the following sequence: the sand pack was saturated with the natural gas (see Table 5.1), which was then filtered using a constant pressure drop until a constant flow rate was established; the sand pack was saturated with n-hexane under vacuum, which was then displaced by natural gas under a constant pressure drop until a constant flow rate was established. The experiment was subsequently repeated in a sand pack that had been treated using an oleophobic agent. The experiment was performed at 25 °C, with a pressure drop of 0.1 MPa ($P_i = 2.1$ MPa, $P_o = 2$ MPa). The absolute permeability of the porous medium was 0.2 µm². The contact angles of quartz wetting by n-hexane were determined before and after treatment by the oleophobic agent. The contact angles were 26° and 108° respectively.

Figure 5.9 shows the experimental results. A significant (almost 20%) flow rate increase occurs in the oleophilic porous medium after n-hexane injection. No such effect was observed in the oleophobic medium, and the gas flow rate after displacement by n-hexane becomes equal to the flow rate before n-hexane injection. Thus, this experiment demonstrates the mechanism of the influence of wettability on the steady-state gas condensate flow.

5.2.3 Mechanism of Pressure Build-Up at the Unsteady-State Flow of the Gas Condensate

Compressibility determines when the pressure build-up occurs. In the considered case, the gas condensate compressibility increases when the pressure decreases (see Figure 5.4). This is related to condensate separation and the associated gas dehydration. Hence, the

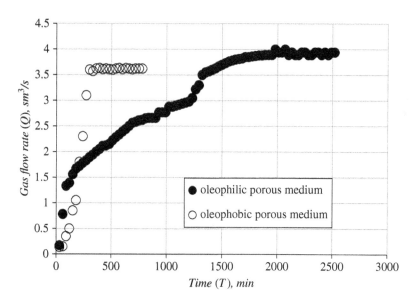

Figure 5.9 Dynamics of the gas flow rate in the oleophilic (the steady-state gas flow rate before the injection of n-hexane was 3.27 cm³/s) and oleophobic (the steady-state gas flow rate before the injection of n-hexane was 3.62 cm³/s) porous media.

calculations provided below assume that only the gas moves in the porous medium after fluid separation. The diffusivity equation in a linear gas flow can be written as follows [62]:

$$\frac{\partial P^2}{\partial t} = \frac{\partial}{\partial x}\left(D\frac{\partial P^2}{\partial x}\right) \tag{5.25}$$

where D is the hydraulic diffusivity of the porous medium and t is time.
The following initial and boundary conditions were used:

$$P^2\big|_{t=0} = P_i^2 - \frac{2\eta P_{at} Q_s x}{kF}; \quad P^2\big|_{x=0} = P_i^2; \quad -\frac{k}{2P_{at}\eta}\frac{\partial P^2}{\partial x}\bigg|_{x=l} \quad F = Q_s\left(1 - \frac{t}{T_s}\right) \tag{5.26}$$

where Q_s is the steady-state gas flow rate at $t = 0$ and T_s is the time at which the inflow terminates during pressure build-up.

The hydraulic diffusivity is inversely proportional to the compressibility. Therefore, it is possible to consider the influence of the compressibility on the pressure build-up process by introducing the dependence of the hydraulic diffusivity on the pressure into Equation (5.25). The former is taken as linear in the first approximation:

$$D = D_o + \frac{D_i - D_o}{P_i - P_o}(P - P_o) \tag{5.27}$$

where D_i and D_o are the hydraulic diffusivities at the porous medium inlet and outlet pressures respectively. Averaging $\partial P/\partial t$ over the length of the porous medium produces

$$\phi(t) = \frac{1}{l}\int_0^l \frac{\partial P^2}{\partial t}dx \tag{5.28}$$

where $\phi(t)$ is a function of t.
Upon inserting Equation (5.28) into Equation (5.25) and after integration,

$$D\frac{\partial P^2}{\partial x} = \phi(t)x + C_1 \tag{5.29}$$

Inserting Equation (5.27) into Equation (5.29) after integration produces

$$D_i P^2 + \frac{D_o - D_i}{P_i - P_o}\left(P_i P^2 - \frac{2}{3}P^3\right) = \phi(t)\frac{x^2}{2} + C_1 x + C_2 \tag{5.30}$$

where C_1 and C_2 are integration constants.
Applying the boundary conditions from Equation (5.26) to Equation (5.30) and accepting the first approximation in which $P^3 = P_i P^2$ produces

$$P^2 = P_i^2 + \frac{\phi(t)}{a}(x^2 - lx) - \frac{2P_i P_{at} Q_s}{Fm\,a}\left(1 - \frac{t}{T_s}\right)x \tag{5.31}$$

where m is the porosity and $a = -\frac{2}{3}\frac{D_i - D_o}{P_i - P_o}P_i + D_o + \frac{D_i - D_o}{P_i - P_o}P_o$

Then, combining Equations (5.28) and (5.31) produces

$$\dot{\phi}(t) + \frac{3a}{l^2}\phi(t) = \frac{3P_i P_{at} Q_s}{Fm\, T_s l} \tag{5.32}$$

Integrating differential Equation (5.32) with the initial conditions in Equation (5.26) produces

$$P^2 = P_i^2 - \frac{2\eta P_{at} Q_s}{kF} x e^{-\frac{3a}{l^2}t} + \frac{2P_i P_{at} Q_s}{Fma} x e^{-\frac{3a}{l^2}t}$$

$$+ \frac{P_i P_{at} Q_s}{Fma T_s} \frac{l\,1}{a}\left(\frac{x^2}{2} - lx\right)\left(1 - e^{-\frac{3a}{l^2}t}\right) - \frac{2P_i P_{at} Q_s}{Fma}\left(1 - \frac{t}{T_s}\right)x \tag{5.33}$$

The pressure at the outlet of the porous medium is determined using Equation (5.33) at $x = l$, and takes the form

$$P_o^2 = P_i^2 - \frac{2\eta P_{at} Q_s l}{kF} e^{-\frac{3a}{l^2}t} + \frac{2P_i P_{at} Q_s l}{Fma} e^{-\frac{3a}{l^2}t}$$

$$- \frac{P_i P_{at} Q_s}{2FmT_s}\frac{l^3}{a^2}\left(1 - e^{-\frac{3a}{l^2}t}\right) - \frac{2P_i P_{at} Q_s}{Fma}\left(1 - \frac{t}{T_s}\right)l \tag{5.34}$$

Figure 5.10 shows the experimental results alongside the numerical results produced using Equation (5.34) and the following parameters: $m = 0.2$, $k = 2 \times 10^{-14} m^2$, $\eta = 1.2 \times 10^{-5}$ Pa s, $l = 1.1$ m, $F = 8.05 \times 10^{-4}\,m^2$, $D_{in} = kP_{in}/\eta m$, $D_o = kP_o/\eta m$, $P_{at} = 10^5$ Pa, $P_i = 27.6$ MPa, $P_o = 26.8$ MPa, $Q_s = 33.3 \times 10^{-6}\,m^{3-1}$, and T_s is taken from the experimental values in Figure 5.3. Figure 5.10 shows that the calculations describe the experimental results quite well.

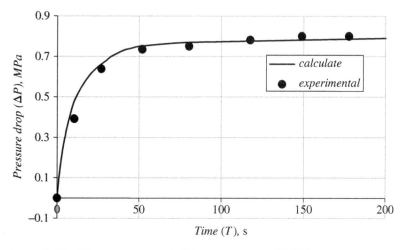

Figure 5.10 Calculated pressure build-up curve at P = 27.2 MPa.

5.2.4 Concluding Remarks

The conclusions of this study are summarized below:

1) The experimental study of gas condensate steady-state flow in the porous medium demonstrated that the gas flow rate started to increase at a pressure that significantly exceeded the dewpoint pressure ($P = 1.74\ P_c$). The gas flow rate reached its peak at $P = 1.5\ P_c$. This peak was almost 30% higher than the flow rate near the critical point. Furthermore, the dependence of the gas flow rate on the pressure was nonmonotonic and the increased flow rates were reached within the $P = 1.4$–$1.74\ P_c$ pressure range.
2) The influence of wettability on the steady-state flow was analyzed. The boundary layer of the condensate does not develop in the oleophobic porous medium and therefore no increase in the gas flow rate is observed.
3) The unsteady-state flow of the gas condensate was studied experimentally. Significant reductions in hydraulic diffusivity and compressibility occur when the pressure decreases.
4) A mechanism that describes flow alteration effects is suggested and is based on the formation of stable subcritical condensate nuclei, the associated slippage effect, and changes in compressibility. The mechanism of subcritical nucleus stabilization by the combination of surface and electrical forces was considered.
5) A developed mathematical model of steady-state gas condensate flow qualitatively describes the nonmonotonic evolution of the gas flow rate and its maximum value.
6) The mathematical model of unsteady-state gas condensate flow accurately describes the experimental results.

Nomenclature

C	volume concentration of gas phase nuclei
C_0	concentration of the condensate at $p = p_c$
C_1, C_2	integration constants
F	cross-sectional area of the porous medium, m^2
K	Boltzmann constant
K_0	absolute permeability of a porous medium, m^2
K_i	relative permeability
$K(P)$	apparent permeability
N	number of molecules in one nucleus
P	external pressure, Pa
P_{at}	atmospheric pressure, Pa
P_0	reservoir pressure, Pa
P_b	pressure in the bubble (or saturation pressure)
P_c	pressure in the nucleus (or dewpoint pressure) or saturation pressure, Pa
P_e	pressure at the outlet of the porous medium or bottomhole pressure, Pa
P_k	drainage area pressure, Pa
P_{st}	pressure stationary element, Pa
P_f	pressure fluctuating element, Pa
P_0	pressure at the inlet of the porous medium, Pa
Q	Poiseuille gas flow rate, sm^3 s^{-1}
Q_0	relative gas flow rate
Q	liquid flow rates with slip
Q_0	liquid flow rates without slip
Q_S	steady-state gas flow rate at $t = 0$, m^3 s^{-1}
R	capillary radius, μm, or the average radius of the pore channel
R_0	radius of the contact line, m
R_k	drainage area radius, m
S_i	degree of saturation
T	temperature, K
T_s	time at which the inflow terminates during pressure build-up, s
V_f	molecular volume, m^3
V_S	flow rate

Nanocolloids for Petroleum Engineering: Fundamentals and Practices, First Edition.
Baghir A. Suleimanov, Elchin F. Veliyev, and Vladimir Vishnyakov.
© 2022 John Wiley & Sons Ltd. Published 2022 by John Wiley & Sons Ltd.

Nomenclature

W_h	work of the heterogeneous nucleation of the gas phase
a_0, k_0	proportionality coefficient, N m^{-2}
b	slip coefficient
e	elementary charge, C
f	fraction of the capillary surface covered with gas nuclei
h	formation thickness
$k(P)$	effective permeability
l	capillary length or length of the porous specimen, m
m	porosity
r_c	well radius, m
r_n	nucleus radius, m
t	time
v_g	volume of gas molecule
v_s	velocity of the liquid slip flow
v_1	velocity, m s^{-1}
v_0	Poiseuille liquid flow velocity, m s^{-1}
v_c	cumulative flow velocity, m s^{-1}
ze	electrical charge on the nucleus surface
ε	dielectric permittivity
ε_0	electrical constant, F m^{-1}
ε_1	dielectric permittivity
$\sigma(r)$	surface tension, mN m^{-1}
σ_0	surface tension of the plane interface
δ	thickness of the nucleus surface layer, m
η_i	fluid viscosity
η_2	viscosity of the gas
μ_i	dynamic viscosity of the fluid
ν	kinematic viscosity
ρ	density of the ball material
ρ_l	liquid density

References

1 Bolotov, A.A., Mirzadzhanzade, A.K., and Nesterov, I.I. (1988). Rheological properties of solutions of gases in a liquid in the saturation pressure zone. *Fluid Dynamics* 23 (1): 143.
2 Ubbelohde, A.R. (1965). *Melting and Crystal Structure*. Oxford: Clarendon Press.
3 Frenkel, J. (1955). *Kinetic Theory of Liquids*. New York: Dover Publications.
4 Zeldovich Ya, B. (2014). *Selected Works. Chemical Physics and Hydrodynamics*. Princeton University Press.
5 Buevich, Y.A. (1987). Precritical nucleus formation in a liquid with surfactant. *Journal of Engineering Physics and Thermophysics* 52 (3): 285.
6 Sirotyuk, M.G. (1970). Elasticity and durability of stable gas bubbles, in water. *Akusticheskij Zhurnal* 16 (4): 567–569.
7 Bunkin, N.F. and Bunkin, F.V. (1992). Bubbstons: stable microscopic gas bubbles in very dilute electrolytic solutions. *Journal of Experimental and Theoretical Physics* 74 (1): 271.
8 Akulitchev, V.A. (1966). Hydratation of ions and the cavitational strength of water. *Akusticheskij Zhurnal* 12: 160.
9 Tyrrell, J.W.G. and Attard, P. (2001). Images of nanobubbles on hydrophobic surfaces and their interactions. *Physics Review Letters* 87: 176104.
10 Simonsen, A.C., Hansen, P.L., and Klosgen, B. (2004). Nanobubbles give evidence of incomplete wetting at a hydrophobic interface. *Journal of Colloid and Interface Science* 273: 291.
11 Agrawal, A. and McKinley, G.H. (2005). *Materials Research Society Symposium Proceedings* 899: 0899-N07-37.
12 Green, H. and Lane, W. (1964). *Particulate Clouds: Dusts, Smokes and Mists*, 2e. London: Van Nostrand, Princeton, N.J.
13 Alty, T. (1926). The origin of the electrical charge on small particles in water. *Proceedings of the Royal Society A* 112: 235–251.
14 Tolman, R.C. (1949). The superficial density of matter at a liquid-vapor boundary. *Journal of Chemical Physics* 17: 118.
15 Navier, C.L.M.H. (1823). Mémoire sur les lois du mouvement des fluides. *Mémoires de l'Académie Royale des Sciences de l'Institut de France* 1: 389–440.
16 Tretheway D., Liu X., and Meinhart C. (2002). Abstracts of papers, 11 Int. Symp. Application of Laser Techniques to Fluid Mechanics. Lisbon.
17 Hirth, J.P. and Pound, G.M. (1963). *Condensation and Vaporization*. Oxford: Pergamon.

18 Zhu, Y. and Granick, S. (2002). Limits of the hydrodynamic no-slip boundary condition. *Physical Review Letters* 88: 106102.

19 Suleimanov, B.A. and Azizov, K.F. (1995). Specific features of the fiow of a gassed liquid in a porous body. *Colloid Journal* 57 (6): 818–823.

20 Kunert, C. and Harting, J. (2008). On the effect of surfactant adsorption and viscosity change on apparent slip in hydrophobic microchannels. *Progress in Computational Fluid Dynamics, An International Journal* 8: 197–205.

21 Esipova, N.E., Zorin, Z.M., and Churaev, N.V. (1981). *Colloid Journal* 43: 9.

22 Suleimanov, B.A., Suleymanov, A.A., Abbasov, E.M., and Baspayev, E.T. (2018). A mechanism for generating the gas slippage effect near the dewpoint pressure in a porous media gas condensate flow. *Journal of Natural Gas Science and Engineering* 53: 237–248.

23 Buevich, Y.A. (1967). On the theory of joint flow of immiscible fluids in a gravity field. *Fluid Dynamics* 2: 113–114.

24 Vinogradova, O.I. (1995). Drainage of a thin liquid film confined between hydrophobic surfaces. *Langmuir* 11 (6): 2213–2220.

25 Vinogradova, O.I., Bunkin, N.E., Churaev, N.V. et al. (1995). Submicrocavity structure of water between hydrophobic and hydrophilic walls as revealed by optical cavitation. *Journal of Colloid and Interface Science* 173 (2): 443–447.

26 Churaev, N.V., Sobolev, V., and D, Somov A.N. (1984). Slippage of liquids over lyophobic solid surfaces. *Journal of Colloid and Interface Science* 97 (2): 574–581.

27 Vinogradova, O.I. (1994). Hydrodynamic interaction between a hydrophobic body and a hydrophilic one. *Colloid Journal* 56 (1): 39–44.

28 Suleimanov, B.A. (2006). *Filtration Features of Heterogeneous Systems*. Izhevsk Institute of Computer Science. ISBN: 5-93972-592-9.

29 Collins, R.E. (1961). *Flow of Fluids through Porous Materials*. New York: Reinhold Pub. Corp.

30 Leonov, E.G. and Isaev, V.I. (2010). *Applied Hydro-Aeromechanics in Oil and Gas Drilling*. Wiley.

31 Suleimanov, B.A. (2011). Mechanism of slip effect in gassed liquid flow. *Colloid Journal* 73 (6): 846.

32 Mikhailov, D.N. and Stepanova, G.S. (2003). Influence of adsorption and desorption of gas micronuclei on the nature of the flow of a gassy liquid through a porous medium. *Fluid Dynamics* 38: 752–759.

33 Bird, R., Stewart, W., and Lightfoot, E. (2002). *Transport Phenomena*, 2e. Wiley.

34 Suleimanov, B.A. (1997). Slip effect during flow of gassed liquid. *Colloid Journal* 59 (6): 749–753.

35 Savins, J.G. (1969). Non-Newtonian flow though a porous industrial medium industrial. *Journal of Industrial and Engineering Chemistry* 61 (10): 18–47.

36 May Van Zi (2003). Emproving development efficiency of the fractured fields on the basis of the dunamic analysis (by the example of the White Tiger field). PhD Dissertation. Baku, ASOA.

37 Shtibina N.B. (1988). Determination of the rocks sorption capacity according to hydrodynamic studies. PhD Thesis. Baku, AzINEFTEKHIM. [in Russian].

38 Pirson, S.J. (1958). *Oil Reservoir Engineering*. New York: McGraw-Hill.

39 Rahuma, K. and Edreder, E. (2013). Models predicting saturated and undersaturated viscosity for some Libyan crude oils. *Petroleum and Coal* 55 (2): 113.

40 Dindoruk B. and Christman P.G. PVT properties and viscosity correlations for Gulf of Mexico oils. SPE-71633-MS. *Presented at the SPE Annual Technical Conference and Exhibition*, New Orleans, Louisiana (30 September–3 October 2001).
41 Whitson, C.H. and Brule, M.R. (2000). *Phase Behavior*. Richardson, Texas: SPE.
42 Danesh, A. (1998). *PVT and Phase Behavior of Petroleum Reservoir Fluids*. Amsterdam: Elsevier Science B.V.
43 McCain, W.D. (1990). *The Properties of Petroleum Fluids*, 2e. Tulsa, Oklahoma: PennWell Publishing Co.
44 Bergman D.F. and Sutton R. P. An update to viscosity correlations for gassaturated crude oils. *SPE-110195-MS. Presented at the SPE Annual Technical Conference and Exhibition, Anaheim, California, USA* (11–14 November 2007).
45 Kahn A. (1987). Development of viscosity correlations for crude oils. *SPE-17132-MS. SPE Publications*.
46 Ahmed, T.H. (1989). Contribution in petroleum geology and engineering. In: *Hydrocarbon Phase Behavior*, vol. 7. Houston, Texas: Gulf Publishing Company.
47 Ahmed, T.H. (2007). *Equations of State and PVT Analysis: Applications for Improved Reservoir Modeling*. Houston: Texas, Gulf Publishing Company.
48 Emets, B.G. (1997). Determination of the average size and concentration of air bubbles in water by nuclear magnetic resonance. *Technical Physics Letters* 23 (7): 513.
49 Suleimanov, B.A., Azizov, F., and Abbasov, E.M. (1998). Specific features of the gas-liquid mixture flow. *Acta Mechanica* 130 (1): 121.
50 Slezkin, N.A. (1955). *Dynamics of Viscous Incompressible Fluids*. Moscow: Gostekhizdat.
51 Volmer, M. (1966). *Kinetics of Phase Formation*. Ohio: Central Air Documents.
52 Hosein, R., Mayrhoo, R., and McCain, W.D. Jr. (2014). Determination and validation of saturation pressure of hydrocarbon systems using extended Y-function. *Journal of Petroleum Science and Engineering* 124: 105–113.
53 Muskat, M. (1946). *The flow of homogeneous fluids through porous media*. Michigan: I.W. Edwards, Inc. Ann Arbor.
54 Enghoff, M.B. and Svensmark, H. (2008). The role of atmospheric ions in aerosol nucleation – a review. *Atmospheric Chemistry and Physics* 8 (16): 4911–4923.
55 Tompkins, A. (2016). *Atmospheric Physics*. Italy, Trieste: ICTP.
56 Klinkenberg, L.J. (1941). *The Permeability of Porous media to Liquids and Gases*. USA: American Petroleum Institute.
57 Lauga, E., Brenner, M.P., and Stone, H.A. (2005). Microfluidics: The no-slip boundary condition. In: *Chapter 15 in Handbook of Experimental Fluid Dynamics* (ed. J. Foss, C. Tropea and A. Yarin). New-York: Springer.
58 Tretheway, D. and Meinhart, C. (2004). A generating mechanism for apparent fluid slip in hydrophobic microchannels. *Physics of Fluids* 16: 1509.
59 Kisileva, O.A., Sobolev, V.D., Starov, V.M., and Shurayev, N.V. (1979). Change in the viscosity of water near quartz surfaces. *Colloid Journal* 41 (2): 245.
60 Lykelma, J. and Overbeek, J.T.G. (1961). On the interpretation of electrokinetic potentials. *Journal of Colloid and Interface Science* 17: 501.
61 Jamiolahmady, M., Danesh, A., Tehrani, D.H., and Duncan, D.B. (2003). Positive effect of flow velocity on gas–condensate relative permeability: Network modelling and comparison with experimental results. *Transport in Porous Media* 52: 159.
62 Bourdarot, G. (1998). *Well Testing: Interpration Methods*. Paris: Edition Techship.

Part C

Production Operations

6

An Overview of Nanocolloid Applications in Production Operations

Nowadays a serious increase in energy demand constrains petroleum engineers to explore new areas to overcome conventional operational challenges. For instance, deep-water operations, high pressure and high temperature (HPHT) formations, and high salinity connate water are only a small part of tasks that could be listed in this regard [1]. Drilling and production are becoming more costly from year to year. Researchers are trying to find more cost effective and profitable ways to improve the performance of well operations. In general, there are only two ways: to find completely new technology/composition/method or to improve existing ones. The last approach is always based on integration of contemporary and traditional science achievements. In this regard, nanotechnology application is no exception [2]. This chapter focuses on utilization of nanocolloids for near well bore production operations.

Well cementing is one of the main and critical well construction processes based on cement injection in the annulus to seal the different formations as well as to provide structural support for the casing. At the same time, cement is also injected to perform oil driving jobs, plugging, well abandonment, and other undertakings. Unfortunately, cement injection does not guarantee that primary aims will be achieved and fluid leakage through the cemented annulus occur quite often. Essential reasons of the mentioned problem could be countless, including inappropriate cementing and unfavorable reservoir conditions (i.e. high connate water mineralization, high reservoir temperature, etc.). The high cost of remedy and production operations encourage researchers to obtain high strength cement formulations in order to avoid any unforeseen complications.

Ozyildirim and Zegetosky showed that nano-SiO_2 and nano-C-S-H (calcium-silicate-hydrate) accelerate the hydration reaction and produce a reduction in wait-on-cement (WOC) time. Furthermore, nanoparticles filled spaces between C-S-H gel lead to porosity and permeability reduction. It was reported that other properties of cement (e.g. porosity, rheology, corrosion resistance, etc.) were also changed. However, not all of them were desirable. The authors mentioned an increase in unwanted viscosity that could lead to pumpability issues and attributed this observation to the high surface area of nanoparticles. Therefore, to achieve the required results suitable optimization should been performed [3].

Jalal et al. presented a study of SiO_2 micro- and nanoparticle impacts on the high performance of self-compacting concrete (HPSCC). The authors show that nanoparticle addition improves the stability of cement slurry by preventing segregation while blending.

Nanocolloids for Petroleum Engineering: Fundamentals and Practices, First Edition.
Baghir A. Suleimanov, Elchin F. Veliyev, and Vladimir Vishnyakov.
© 2022 John Wiley & Sons Ltd. Published 2022 by John Wiley & Sons Ltd.

The following considerations have been made: paste volume, resistivity, and compressive strength were increased and water absorption and capillary absorption were reduced [4].

Researchers mainly focused on SiO_2 nanoparticle application due to a convenient manufacturing process and well-known chemical properties. The last but not the least reason is an environmentally friendly property of SiO_2 nanoparticles and the possibility to produce them with various wettability behaviors (i.e. hydrophilic and hydrophobic) [5].

In summary, cement property enhancement attributed to the nanosilica addition could be listed as follows:

1) Bonding increase between aggregates and cement paste [6].
2) Compressive strength increase [3, 7–9].
3) Permeability decrease [3, 4, 9].
4) Improved microstructure [3, 4, 9].

Qian et al. reported that cement demonstrates a self-healing behavior with nanoclay addition. The mechanism of the observed phenomena is based on a water-retaining capacity of nanoclay and nanoclay acting as an internal water source for further hydration [10].

Santra et al. presented a study of the impact of several nano-additives on the well cementing process. The authors declared that nanosilica and nanoalumina particles act as accelerators in cementing slurry. Carbon nanotubes lead to the enhancement of mechanical properties. The improvement was also observed in cement structures and an increment in thermal/electrical conductivity [11].

It should be noted that in terms of reinforcing materials for cement, carbon nanotubes were found to be the most suitable [12].

Maserati et al. proposed the use of nanoemulsions as the spacer in cementing operations. The authors observed an improvement of an open hole cleaning capacity as well as a wettability alteration that leads to better slurry adhesion in the annulus. One of the main advantages of this technology is the reduction in chemical consumption. The latter application makes the cementing operation more cost effective [13].

Nanoparticles (NPs) are also extensively used to moderate the drilling fluid properties (e.g. rheology and filtration behavior). The authors obtained two patents in the area of NP application for drilling fluids. First of all, it was stated that both low and high concentrations of NPs in the drilling fluid show almost equal performance. In this regard it is not economically profitable to utilize high concentration of NPs [14, 15]. In addition, there is a direct dependence between an increase in NP concentration and the friction coefficient leading to a hole cleaning efficiency reduction with high NP concentrations [16]. Some NP even showed a negative impact on filtration characteristics at high temperature situations. Copper oxide nanoparticles have a better thermal stability compared to other types of NP. Minakov et al. reported that silicon dioxide NP increases pressure losses in water-based drilling fluids. The authors attributed the observed phenomena to the frictional forces increment [17].

Amanullah and Al-Tahini investigated the impact of nanomaterial-based drilling fluid on the drilling bit and stabilizer ballbearings. It was found that nano-drilling fluid high surface area prevents sticking of the highly reactive shale to the bit and tool joints, consequently preventing the reduction of penetration rate (ROP). In the meantime, fine and very thin film formed by NPs reduces frictional resistance between the formation and pipe.

This phenomenon makes nano-based drilling fluids a prospective solution for drag and torque issues during horizontal, coiled tubing, and multilateral drilling [18, 19].

Price et al. declared that nanoparticles can provide an effective solution for oil-based mud applications in the environmentally sensitive areas [20].

Sayyadnejad et al. reported that ZnO_2 (14–25 nm) particles act as very effective agents for hydrogen sulfide removal from water-based drilling fluids. The authors stated that addition of zinc oxide nanoparticles is able to completely remove hydrogen from drilling mud in 15 minutes [21].

Nanoparticle applications were widely investigated in all areas of petroleum production and could be considered as a prospective solution for conventional tasks. The range of applications involved hydrate recovery, scale inhibition, sand control operations, and others is being actively used for nanoparticle applications [22–24]. The following chapter will describe the main impact mechanisms of nanoadditive formulations on colloidal systems used in petroleum production.

7

Nanosol for Well Completion

The increase in the quality of annular space isolation is one of the important tasks both in the construction of wells and during the operational period. One of the most critical factors affecting both areas is the cement stone strength. The ways for improving this property can be traditionally divided into two main groups. The first group includes cements obtained by careful consideration of the cement powder chemical composition. Then there is the use of various fillers and reinforcing elements as additives for the existing product (cement powder).

The positive effect of nano-fillers on the cement stone strength is explained by the increase in crystallization of the hydrated cement particles as a result of their addition [25, 26]. For example, Tao [27], having investigated the microstructure of the cement stone containing nano-SiO_2 using Scanning Electron Microscopy, observed an improvement in the porous structure of the cement stone and the material's denser structure. Works [28, 29] also demonstrated a positive effect of nano-SiO_2 on the cement mechanical properties compared to microsilica. Li and Xiao reported that the solutions containing nano-SiO_2 or nano-Fe_2O_3 have a greater compressive and flexural strength than an ordinary cement mortar [30]. Another study by Li et al. [31] showed a great abrasion resistance of the samples containing nano-SiO_2 and nano-TiO_2.

In the works of Nazari et al. [32–35] it was shown that the addition of TiO_2 nanoparticles up to 4 mass% in cement powder improves mechanical properties of the cement stone.

The second group includes the cements obtained by changing the physical characteristics of the cement powder. In this case, the granulometric composition of the cement powder is altered by increasing the fineness of the grinding.

These changes will be expressed qualitatively in the following changes: size distribution of cement powder particles and specific surface area.

Works [36–38] considered the improvement in the cement stone strength by controlling the granulometric composition, specific surface area, and distribution of particles in the cement powder.

In his work, Celik pointed out the importance of the particle distribution interval and fineness of grinding, especially for the early stages of the cement stone hydration [39]. Bentz et al. reported the results of laboratory studies and computer simulation of the influence of the fineness of cement grinding on the physical and mechanical properties of cement stone, showing a significant improvement in these properties [40].

Nanocolloids for Petroleum Engineering: Fundamentals and Practices, First Edition.
Baghir A. Suleimanov, Elchin F. Veliyev, and Vladimir Vishnyakov.
© 2022 John Wiley & Sons Ltd. Published 2022 by John Wiley & Sons Ltd.

Only the use of nano-additives among the aforementioned methods for increasing the cement stone strength is not accompanied by a change in the production process. All the remaining methods are possible only when revising the technological process of production, which, of course, will affect the cost of the final product (i.e. it is not going to be cheaper). On this basis it is important to justify and optimize the selection of oil well cement from a practical and economic point of view. Mechanisms for increasing the cement stone strength based on the example of Portland cement with nano-TiO_2 (15 nm) and nano-SiO_2 (15 nm) additives, as well as its various finely dispersed modifications obtained in the laboratory, were studied and compared. Two regression equations for the initial and late periods of cement hardening have been obtained.

The starting material for conducting the laboratory tests was Portland cement, which was used to prepare two groups of the cement powder samples:

- The first group of seven samples was obtained by adding TiO_2 and SiO_2 nanoparticles in the quantity from 0.5 to 2 mass% (Table 7.1).
- The second group of five samples was obtained by grinding. For each sample of this group, the specific surface area and granulometric composition (Table 7.2) were measured. X-ray analysis was undertaken and confirmed the identity of the nanopowder structural and phase composition (Table 7.3).

Table 7.1 Concentration of TiO_2 and SiO_2 nanoparticle additives.

Sample	Nano-TiO_2 (mass%)	Nano-SiO_2 (mass%)
S-1	0	0.5
S-2	0	1
S-3	0	1.5
T-1	0.5	0
T-2	1	0
T-3	1.5	0
T-4	2	0

Table 7.2 Granulometric composition and specific surface area of the obtained samples.

Sample	D10	D50	D90	S_{sp} (sm^2 g^{-1})
P-1	1.76	40	62	2630
P-2	1.88	16.6	54	2850
P-3	1.63	16.3	50.6	3640
P-4	1.25	10.4	38.5	4320
P-5	1.85	17.8	27	4710
P-6	1.98	6.6	16.9	5480

Area conversion factor: $1 \text{ cm}^2 \text{ g}^{-1} = 48.8242 \text{ ft lb}^{-1}$.

Table 7.3 X-ray diffraction analysis.

Phase	C3S	C3S(beta)	C3A	C4AF	Ca(OH)$_2$	MgO	SiO$_2$	K$_2$SO$_4$
% mass	61	18	1.3	14.5	0.3	1.3	0.3	0.5

The water–cement ratio when preparing the solution was 0.44 and no other additives were used.

7.1 The Influence of the Specific Surface Area and Distribution of Particles on the Cement Stone Strength

The data obtained (Table 7.4) show that the specific surface area is poorly correlated with the increase in the cement strength on the first day of hardening. However, after 28 days, this strength is almost directly proportional in all samples.

However, the specific surface area cannot be considered as an absolute criterion for predicting the cement stone strength. In view of the fact that the cements have completely different surface and fractional compositions, they can have equal strengths values. The role of the different-sized fractions is not identical in the process of cement hardening. Thus, it is shown [38, 41] that particles with the size of 0–5 μm contribute to the gain in strength during the first 24 hours, i.e. the thinner the cement is ground, the faster it reacts with water (hydrates) and hardens. The high hydration activity of the fine cement fractions is directly related to their large specific surface area and small relative thickness of the protective layers that appear on the cement grains during hydration. The increase in the strength with an increase in the portion of the finely dispersed fractions is also nonlinear [42]. The cement stone strength in this case grows fast during the first day of hardening and then the growth rate decreases. This is explained by the fact that the amount of nonhydrated particles from larger fractions in the finely dispersed cement is small and their contribution to the gain in

Table 7.4 The influence of the specific surface area and distribution of particles on the cement stone strength.

Sample	1 day Strength (MPa)	2 days Strength (MPa)	7 days Strength (MPa)	28 days Strength (MPa)	D10	D50	D90	S_{sp} (sm^2 g^{-1})
P-1	13.2	28	45.5	56.7	1.76	40	62	2630
P-2	24	30	48.6	62	1.88	16.6	54	2850
P-3	15.3	36.7	53.5	65.2	1.63	16.3	50.6	3640
P-4	12.2	32	61.5	66.2	1.25	10.4	38.5	4320
P-5	22.7	54.28	55.9	71.4	1.85	17.8	27	4710
P-6	28	45	68	81	1.98	6.6	16.9	5480

Conversion factors: 1 MPa = 145.03 773 773 psi and 1 cm^2 g^{-1} = 48.8242 ft^{2}lb^{-1}.

strength is not significant at the later stages. Due to this process, most of the maximum possible strength of this cement can be already observed in the early stages of hardening and the further gain in strength is not as significant. However, the opposite situation when the cement consists mainly of larger fractions is not desirable. In this case, the hydration process proceeds much more slowly, as well as the gain in strength does in its turn.

The results of the experiments show an increase in the strength of the finest-dispersed cement stone (P-6) by 112% and 42% in comparison with the ordinary cement powder (P-1) at the early and the late stages of hardening respectively.

In conclusion, the optimal fractional composition is critical for obtaining a desired cement stone within the required time limits.

7.2 The Influence of Nano-SiO$_2$ and Nano-TiO$_2$ on the Cement Stone Strength

To study the influence of nanoparticles on the cement stone strength, various samples with various amounts of SiO$_2$ and TiO$_2$ nanoparticles were used (Table 7.5). The maximum concentration was 1.5 mass% for nano-SiO$_2$ since the further increase in the concentration of up to 2 mass% resulted in unstable rheological parameters of the cement solution that did not allow it to be used when carrying out the strengthening or isolation operations. However, it should be noted that in the works of other researchers, higher concentrations of nano-SiO$_2$ up to 10 mass% were used, which is most likely connected with different particle sizes (5–70 nm), their distribution, and surface area [43].

Table 7.5 The influence of nanoparticles on the cement stone strength.

Sample	1 day Strength (MPa)	2 days Strength (MPa)	7 days Strength (MPa)	28 days Strength (MPa)	Nanoparticles (% mass)
Nano-SiO$_2$ (% mass)					
P-1	13.2	28	45.5	56.7	0
S-1	13.7	29.2	46.1	58.2	0.5
S-2	17	38	60	67	1
S-3	14	32.8	51.4	64	1.5
Nano-TiO$_2$ (% mass)					
T-1	14.1	31.5	52.3	67	0.5
T-2	14.6	32.9	54.4	64.5	1
T-3	14.9	36.2	58.4	66.6	1.5
T-4	15.4	38.4	59.8	68.1	2

Conversion factor: 1 MPa = 145.03 773 773 psi.

In the case of nano-TiO$_2$, the maximum concentration was 2 mass% due to the fact that the further increase in the cement stone strength was insignificant with an increase in the particle concentration.

It can be seen that the use of SiO$_2$ and TiO$_2$ nanoparticles resulted in an increase in the cement strength by 35.71% and 37.14% in the early stages of hardening and by 18% and 20% respectively at the later stage.

Observed phenomena are explained by the results of the study carried out by Sobolev and Sanchez [43]. They studied the effect of nano-SiO$_2$ on the cement and showed an increase in the rate of its hydration due to directional formation of ordered supramolecular structures of calcium silicate hydrates that penetrate the stone matrix and increase its strength due to a high specific surface area. This effect is connected with high chemical activity of nano-SiO$_2$ that accelerates the hydration reaction, which increases the amount of CHS bonds in the cement paste and forms longer silicate chains.

A proposed statement was verified by SEM investigations of the cement stone (see Figure 7.1). The fibrous structure of the formed cement stone is distinguished when there is nano-SiO$_2$ in the system when compared to the usual Portland cement.

TiO$_2$ nanoparticles, in contrast, appear to act solely due to an increase in the specific surface area, creating a more uniform matrix and reducing the volume of large pores.

Thus, the strength indexes obtained are higher than those for P1–P4 samples and are comparable with the results for the P-5 sample, i.e. with the decrease in the average diameter of Portland cement particles by 56%.

7.3 Regression Equation

In order to determine the existence of a mathematical relationship between the results obtained and identify the possibility of using these results for prediction of the cement stone strength at different stages of hardening, the regression analysis method was used. It should be noted that this method was not applied to the results obtained when carrying out the study of the influence of nanoparticles on the cement strength since there were not enough data to construct a statistically significant regression equation [44].

When constructing the equation, the resulting feature was the value of the cement strength and the factorial features included the following: fractional composition, specific surface area, and hardening time. The regression equation simultaneously describing both stages of the gain in strength is statically unreliable due to the inclusion of the factorial features of various cement fractions, which have an uneven influence on the gain in strength in different stages of hardening.

In this regard, the regression equation was developed for two particular cases:

1) The multiple regression equation for the initial stage of the gain in strength (1.2 days):

$$y = -45.2959 + 16.1208x_1 + 0.0046x_2 + 0.7679x_3$$

$$y = -5.06x_1 + 1.22 \times 10^{-1}x_2 - 1.01x_3 - 8.38 \times 10^{-3}x_4 + 18.38x_5 + 82.78$$

where x_1 is D10, x_2 is surface area, and x_3 is hardening time in days.

The coefficient of determination: $R^2 = 0.925\,343$

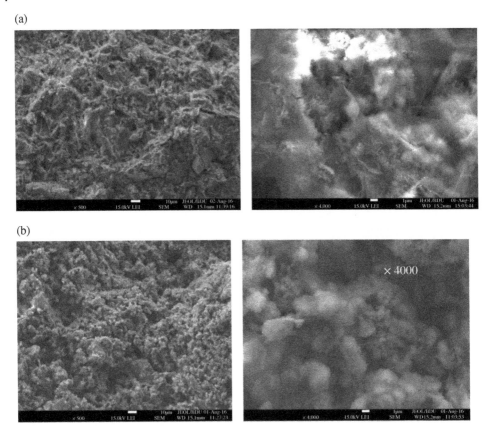

Figure 7.1 SEM images. (a) Portland cement with 1% mass nano-SiO$_2$. (b) Portland cement without nanoparticle additives.

2) The multiple regression equation for the late stage of the gain in strength (7.28 days):

$$y = -0.14437x_1 + 0.006022x_2 + 0.551587x_3 + 30.51363$$

where x_1 is D50, x_2 is surface area, and x_3 is hardening time in days.

The coefficient of determination: $R^2 = 0.972\,884$

The multiple regression equations obtained have a high determination coefficient for the early and late stages of the gain in strength of $R^2 = 0.92$ and $R^2 = 0.95$ respectively, and are statically significant.

7.4 Concluding Remarks

1) The specific surface area cannot be considered as an accurate criterion for predicting the cement stone strength. However, the introduction of such an indicator as the particle size distribution makes it possible to make more accurate predictions.

2) The use of SiO_2 and TiO_2 nanoparticles led to an increase in the cement strength in the early stages of hardening by 35.71% and 37.14%, as well as by 18% and 20% at the later stage, respectively.
3) The multiple regression equations based on the calculation of the specific surface area and particle distribution were obtained. The resulting multiple regression equations have a high determination coefficient for the early and late stages of the gain in strength $R^2 = 0.92$ and $R^2 = 0.95$ respectively, and are statically significant.
4) SiO_2 and TiO_2 nanoparticles generally improve the isolating properties of the cement stone, allowing improvement in the characteristics of conventional Portland cement and are comparable to the decrease in the average diameter of Portland cement particles by 56%.

8

Nanogas Emulsion for Sand Control

8.1 Fluidization by Gasified Fluids

The formations composed of soft and poorly consolidated rocks represent the development challenges as the formation matrix breaks down in the bottomhole zone. Moreover, in this case, the fluid and gas, when moving along the formation, carry some amounts (which is sometimes very significant) of sand into the well. If the flow velocity is not high enough to lift the sand grains, they precipitate on the bottomhole, gather in masses, and form a plug, which can be several hundreds of meters high. Notwithstanding the fact that the sand plug permeability is 10 times higher than the formation permeability, its development results in a well operation stoppage, which is explained by the higher hydraulic resistances of the fluid while filtering through the plug compared with tubing. Water is always present in the production well and this leads to an increase in the pressure gradient and, as a result, to a further formation matrix breakdown. The smaller sand and clay particles start coagulating and precipitating when there is water inflow. This process inevitably leads to a sand plug formation.

In general, sand plug washout is carried out with the help of technical water in the wells at the formation pressure. In these conditions, it is not possible to remove the whole sand plug by just water washing. Some sand portion gets into the formation together with water because of strong absorption, which causes partial bottomhole zone breakdown, and can sometimes lead to production line deformation. Furthermore, during the process of well development, the sand, which penetrated into the bottomhole zone, can freely get into the wellbore and, after some time, it becomes a necessity to remove the sand plug again.

Universal application of water for sand plug removal in production wells is the cause of decommissioning of the producing well stock, as well as of a considerable reduction in current flow rates and the final oil recovery coefficient. Hence, the physical and chemical methods for sand plug washing are the most widespread in field conditions. In particular, the method for sand plug washing with the help of the gas–fluid systems [45] and aqueous solutions of chemical reagents [46] are used internationally. The shortcoming of these methods is the high consumption of chemical reagents and gas. Moreover, when the washup fluids get into the formation, due to their high gas saturation and viscosity, a reduction in gas permeability for oil is possible.

The working fluid for the sand plug washing should have high sand removal properties, which is usually provided by the fluid high viscosity. The working fluid should have low

Nanocolloids for Petroleum Engineering: Fundamentals and Practices, First Edition.
Baghir A. Suleimanov, Elchin F. Veliyev, and Vladimir Vishnyakov.
© 2022 John Wiley & Sons Ltd. Published 2022 by John Wiley & Sons Ltd.

density even despite the fact that the low density decreases the granular material removal velocity. The mutually exclusive requirements for the working fluid causes technical controversy, which is peculiar to the sand plug washing process. The following text investigates the fluidization by heterogeneous systems as a possible way to develop more economical techniques for sand plug washing.

8.1.1 Carbon Dioxide Gasified Water as Fluidizing Agent

The fluidization process of the sand systems was studied on the setup shown in Figure 8.1. The installation incorporated the following elements: an organic glass column (1) with the inner diameter of 1.2×10^{-2} and length of 2.4 m, at the bottom of which is installed a metal filter with the diameter of 0.03×10^{-3} m holes; a saturation tank with the mixer, whose working volume is 5×10^{-2} m^3 (2); a differential pressure gauge (3); block valves (4); a compressed gas cylinder (5); a pressure reducer (6); a standard pressure gauge (7); measuring vessel (8); and a pressure controller (9).

The purpose of the experiments was to study the process of quartz sand fluidization with the help of the water gassed by carbon dioxide at pressures much higher and near the saturation pressure. The saturation pressure P_c was equal to 0.03 MPa in all the experiments. The sand plug model was made of quartz sand of two fractions: coarse sand with an average size of 0.25×10^{-3} m and fine sand with an average size of 0.1×10^{-3} m.

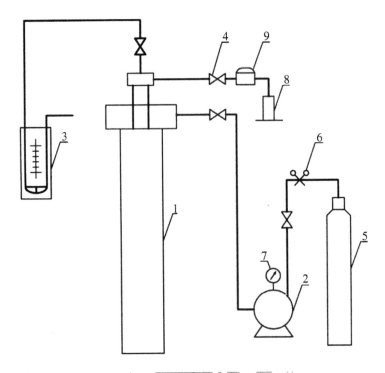

Figure 8.1 Experimental setup.

The experiments were conducted according to the following scheme:

- The organic glass tube was filled with a portion of the preliminary soaked quartz sand.
- The saturation tank was used to prepare the gassed mixture with the help of tap water and carbon dioxide;
- The gasified water was filtered through the granular layer with the help of the compressed gas cylinder and saturation tank.
- During the process of the gassed fluid filtration, the characteristics of fluidization ($\Delta P - Q$) were measured and dependence of the granular layer expansion (Δh) on the fluid flow rate at constant average pressure (P) in the experimental tube was recorded. The average pressure level was also recorded.

The fluidization process was carried out until the sand was removed from the granular layer. To make comparison of the results possible, an experiment was conducted that used degassed water as the fluidizing agent. The experiments were carried out at different average pressures ($P = 0.02$–0.08 MPa). In the first series of experiments a homogeneous sand plug was used (i.e. one made from one sand fraction). The filling height was 0.2 m.

Figure 8.2 shows the dependence of the relative granular layer expansion ($\Delta h_0 = \Delta h_1 / \Delta h_2$, where Δh_1, Δh_2 – granular layer expansion at the studied average pressure and fluidization by degassed water respectively) on the fluid flow rate at $P = 1.5 P_c$. Figure 8.3 shows characteristics of fluidization for degassed and gasified fluids.

It can be seen from Figure 8.2 that the maximum relative granular layer expansion is equal to 2.5, while Figure 8.3 shows that before the fluidization begins, the fluid flow rate for the gasified fluid is higher than the one for the degassed fluid at the same differential pressure. However, in the first case, fluidization begins at the 10% lower differential pressure.

The second series of experiments were conducted at the heterogeneous sand plug (i.e. made from various sand fractions). The filling height was 0.2 m.

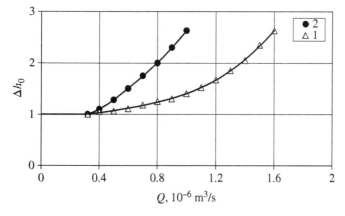

Figure 8.2 Dependence of relative granular layer expansion on the fluid flow rate ($P = 1.5P_c$) for homogeneous (1) and heterogeneous sand plugs (2).

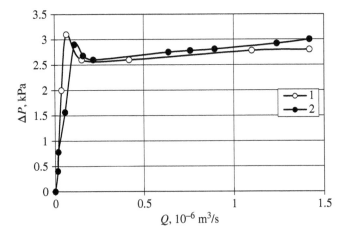

Figure 8.3 Characteristics of sand plug fluidization by degassed water (1) and system water – carbon dioxide at $P = 1.5P_c$ (2).

It can be seen from Figure 8.2 that much smaller fluid flow rate values correspond to the same values of the relative granular layer expansion for the heterogeneous plug. The flow rate, which corresponds to the granular material removal from the fluidized layer, is 40% lower for the heterogeneous plug than for the homogeneous plug. Furthermore, the flow rate characteristics turned out to be the same, as shown in Figure 8.3.

Figure 8.4 shows the dependence of the granular material expansion on the average pressure level at a constant fluid flow rate ($Q = 1.4 \times 10^{-6}$ m^3 s^{-1}). It can be seen from Figure 8.4 that the granular material expansion increases in proportion to the average working pressure approximation of the saturation pressure. The maximum value of the granular layer expansion occurs at $P = 0.045$ MPa ($P = 1.5P_c$). When the pressure decreases below the saturation pressure, i.e. during the two-phase system flow, the granular layer expansion is lower compared to the degassed water application (or at $P > 2P_c$).

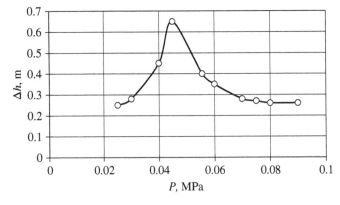

Figure 8.4 Dependence of granular layer expansion at a constant fluid flow rate on average pressure for a system of water and carbon dioxide.

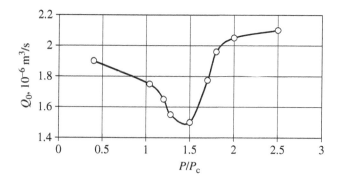

Figure 8.5 Dependence of the fluid flow rate that corresponds to the beginning of granular material removal from the fluidized layer on the relative pressure level for the water–carbon dioxide system.

Figure 8.5 shows the dependence of the fluid flow rate, which corresponds to the beginning of the granular layer removal from the fluidized layer Q_0, on the relative average pressure level (P/P_c). It can be seen from the figure that the specified dependence is of a nonmonotonic nature, with the minimum at $P = 1.5P_c$. Furthermore, Q_0 decreases by almost 30% when compared to $P = 1.5P_c$.

Thus, application of the gasified fluid in the pre-transition phase region can significantly improve removal of the sand plug. Particularly, it is more efficient in the case of the heterogeneous sand plug.

8.1.2 Natural Gas or Air Gasified Water as Fluidizing Agent

It is commonly known that one of the main methods for oil production is the compressor method since it is much cheaper to use water mixtures with air or natural gas as an agent for sand plug washing in field conditions. To determine the efficiency of washing with these mixtures, laboratory experiments were carried out with the help of the installation shown in Figure 8.1. The only difference was that the experimental column diameter was equal to 2.4×10^{-2} m. In both cases, the filling height of the sand was equal to 0.2 m and the saturation pressure was 0.2 MPa. A heterogeneous plug was modeled in the experiments.

The results of the experiments are shown in Figures 8.6–8.9. It can be seen from Figures 8.6 and 8.7 that the granular layer expansion increase is observed in the region of intense nucleation pressures. Moreover, the maximum effect in both cases is achieved at $P = 0.25$ MPa (i.e. at $P = 1.25P_c$). Furthermore, the relative granular layer expansion for the water–air mixture is at around 40% and is approximately 30% for the water–gas mixture (see Figures 8.8 and 8.9).

It should be noted that the fluid flow rate while application was taking place for water–air and water–gas mixtures was 20–30% lower compared to degassed water. Utilization of water–air and water–gas mixtures as fluidization agents in the pre-transition phase region significantly improved sand plug removal compared to the degassed water.

150 | 8 Nanogas Emulsion for Sand Control

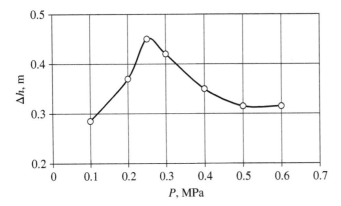

Figure 8.6 Dependence of granular layer expansion at a constant fluid flow rate on the pressure level for the water–air system.

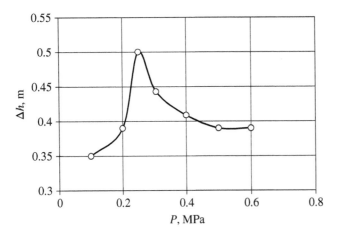

Figure 8.7 Dependence of granular layer expansion at a constant fluid flow rate on the pressure level for the water–natural gas system.

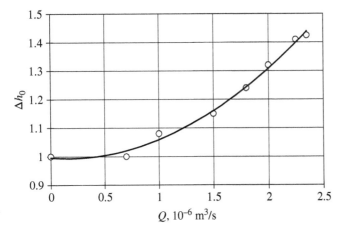

Figure 8.8 Dependence of relative granular layer expansion on the fluid flow rate for the water–air system.

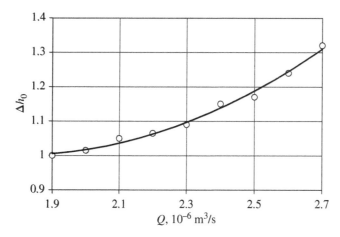

Figure 8.9 Dependence of relative granular layer expansion on the fluid flow rate for the water–natural gas system.

8.2 Chemical Additives Impact on the Fluidization Process

8.2.1 Water–Air Mixtures with Surfactant Additives as Fluidizing Agent

The experiments were carried out on the setup shown in Figure 8.1. The experiments differed from the ones conducted before, as the water–air mixture ($P_c = 0.2$ MPa) with different surfactant additives was supplied to the column inlet.

An anionic surfactant (sulphanole) was used as an additive in the first series of experiments. The results are shown in Figures 8.10 and 8.11. It can be seen from Figure 8.10 that when the surfactant is added, the relative granular layer expansion decreases. However, at $C < 0.05\%$, it is higher than the one for the degassed water. It appears from Figure 8.11 that the dependence of the granular layer expansion on the surfactant concentration has a nonmonotonic nature with the minimum at the point $C = 0.05\%$.

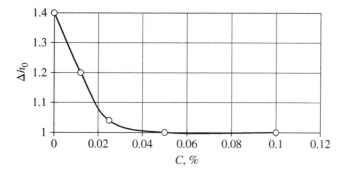

Figure 8.10 Dependence of relative granular layer expansion on the anionic surfactant concentration.

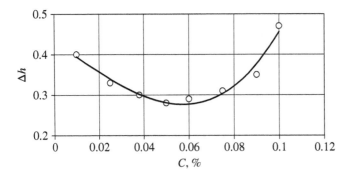

Figure 8.11 Dependence of granular layer expansion on the anionic surfactant concentration.

The cationic surfactant (catapine) was used as an additive in the second series of experiments. The results are shown in Figure 8.12, where it can be seen that, when the cationic surfactant is added, the relative granular layer expansion increases monotonously and, at $C = 0.05\%$, it reaches the value of 2.1, which is 50% higher than its value without surfactant additives.

8.2.2 Fluidization by Polymer Compositions

The experiments were conducted on the setup shown in Figure 8.1. However, a polymer composition (0.02% PAA, 0.005% sulphanol), and a water–air mixture respectively were supplied in the pre-transition phase state to the column inlet. The results are shown in Figure 8.13, where it can be seen that, at s constant flow rate $Q = 3.8 \times 10^{-6}$ m^3 s^{-1}, the granular layer expansion is 15 and 32% higher for the polymer composition and the water–air mixture respectively compared to the degassed water. The fluid flow rate in the pre-transition phase corresponding to the sand removal from the fluidized layer is 10–15% lower for the water–air mixture in comparison with the polymer composition.

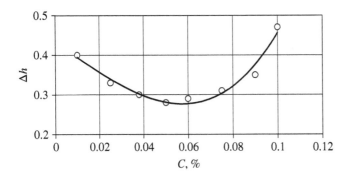

Figure 8.12 Dependence of relative granular layer expansion on cationic surfactant concentration.

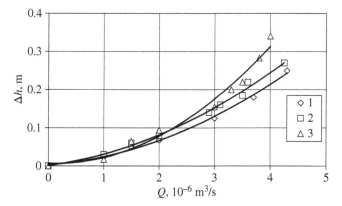

Figure 8.13 Dependence of granular layer expansion on the fluid flow rate: 1 – water, 2 – polymer composition, 3 – water-air mixture at $P = 1.25P_c$.

8.3 Mechanism of Observed Phenomena

To compare the experiments of different diameter columns, the dependence of the relative granular material expansion on the filtration velocity for different mixtures was plotted. Figure 8.14 shows the specified dependence, which was plotted in asymptotic coordinates ($Y = h_0/h_{0m}$, $X = v/v_m$, where h_{0m} and v_m are maximum values of the granular layer expansion and flow velocity respectively) and from which it can be seen that the input data in the asymptotic coordinates make a single curve that is indicative of the qualitative unity of processes during fluidization of the studied gassed fluids.

It should be noted that the maximum value Δh_0 was for carbon dioxide at $P = 1.5P_c$ and for air and natural gas at $P = 1.25P_c$. At $P = 1.5P_c$, the value Δh_0 is distributed as follows: 1.45 for carbon dioxide; 1.35 for air; and 1.15 for natural gas. The maximum value is connected with different solubilities of the specified gases in water. Furthermore, at normal conditions, the carbon dioxide solubility is much higher than the air or natural gas solubility in water. Therefore, in equal conditions, the fluid gas saturation in the case of carbon

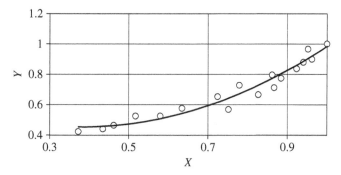

Figure 8.14 Dependence of relative granular layer expansion on flow velocity in asymptotic coordinates.

dioxide saturation is much higher than the saturation of air and natural gas, although air saturation is higher than natural gas saturation (used in the experiments). The volume concentration of the formed gas phase nuclei increases when the fluid gas saturation increases, which determines the growth of the relative granular layer expansion value when gas solubility increases. Furthermore, in this case, when the fluid gas saturation reduces, the effect peak approaches the saturation pressure (the granular layer expansion maximum is achieved at $P = 1.5P_c$ for carbon dioxide and at $P = 1.25P_c$ for air and natural gas).

Richardson and Zaki proposed an empiric relation between the fluidization velocity (U) and fluidized layer porosity (ε), and the linear dependence in logarithmic coordinates, which became widely accepted as [47]

$$\frac{U}{U_i} = \varepsilon^n$$

where U_i is the suspension settling velocity.

Figure 8.15 shows the dependence of the fluidization velocity (10^{-3} m s^{-1}) on fluidized layer porosity (%), plotted in logarithmic coordinates. The curves belong to degassed fluid (water) (curve 1) and carbon dioxide gassed water (curve 2).

It can be seen from the figure that fluidization with degassed water is described by the Richardson–Zaki equation, while there are two linear sections on the presented dependence when fluidization is carried out with a gasified fluid.

Considering the above, the fluidization process can be described by the following generalized equation:

$$U = \sum_{i=1}^{k} a_i \exp(-b_i \varepsilon)$$

The fluidization process of a gassed fluid is described by this equation with $k = 2$. The experimental results are satisfactorily described at the following values of the constant coefficients: $a_1 = 964.29$; $a_2 = 15.02$; $b_1 = 0.107$; $b_2 = 0.046$. The fluidization process of the degassed fluid is described by this equation at $k = 1$. The experimental results are satisfactorily described at the following values of the constant coefficients: $a = 0.0379$; $b = -0.0576$.

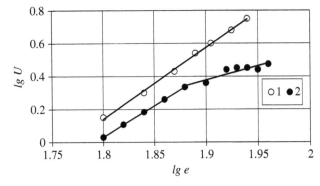

Figure 8.15 Dependence of fluidization velocity on granular layer porosity for degassed (1) and gassed fluid (2) in logarithmic coordinates.

8.3 Mechanism of Observed Phenomena

Based on the experimental data, a possible observed phenomena kinetic mechanism can be proposed. It is known that the provided experimental results for the fluid fluidization do not align well with the theoretical model. The real permeability of the fluidized layer turns out to be higher than the theoretical one, which is explained by agglomeration of particles into small groups (coils or floccules) [48]. At the same time, in the work of Zhou and coworkers [49], it is shown that the nuclei of the gas phase (air) with submicron sizes, when forming from water on to the surface of the granular material particles (silica), form a unique "solvation shell" that prevents further granule adhesion. It is natural to assume that, in this case, the gas nuclei, when attaching to the surface of sand particles, prevent further adhesion between particles and thus the granular material fluidization accelerates. Furthermore, considering the fact that the aggregation of particles significantly increases when the liquefaction velocity increases (due to more frequent particle collisions) [48], the second linear section with lower n value appears on the dependence $U(\varepsilon)$. This is indicative of the average size reduction of the sand particles [47, 48] due to suppression of adhesion (lower agglomeration). It should be noted that the fluid slippage influence is insignificant in the analyzed case. In accordance with reference [50], when sedimentation occurs in the lyophobic submicron quartz particles, the slippage effect can increase the sedimentation velocity (or suspension velocity) by 20%; moreover, the effect decreases sharply when the sizes of particles increase. According to reference [50],

$$\frac{v_s}{v_{ns}} = \frac{1 + \dfrac{3b}{R}}{1 + \dfrac{2b}{R}}$$

where v_s, v_{ns} are sedimentation velocities with slippage and without it respectively, b is the slippage coefficient, and R is the particle radius. For the particles with the sizes of 10^{-4} and 2.5×10^{-4} m that were used in the experiment, at $b \approx 10^{-6}$ m [50], the estimations in accordance with the specified formula give $v_s/v_{ns} = 1.003 - 1.009$, i.e. the sedimentation velocity, increases by only 0.3–0.9%.

The anionic surfactant additive, due to its adsorption on the particles surface, improves wettability and decreases volume concentration of nucleation (see reference [51]). This, in turn, reduces the number of grains covered by the "salvation shell." An increase of Δh_0 at $C > 0.05\%$ is connected with the grain micellization and accompanying increase in the working agent viscosity. The cationic surfactant additive and accompanying hydrophobization of the granular material surface, on the contrary, promote an increase of the volume concentration (see reference [51]) and the number of sand particles covered by the "salvation shell." Sand removal worsens in the anionic case and improves in the cationic case. The real sand plugs are usually hydrophobized by the hydrocarbon fluid so they will be better removed by gasified fluids in the pre-transition phase state according to the obtained results.

It is possible to conclude that application of the gasified fluids in the pre-transition phase state (including the ones with the surfactant additives) increases the washing-over process efficiency compared to the cases when the degassed fluids and two-phase gas–fluid mixtures are used.

9

Vibrowave Stimulation Impact on Nanogas Emulsion Flow

An efficient method of raising well productivity is vibrowave action on the field and the critical area. For this purpose a vibration generator needs to be installed. The generator creates elastic waves acting on the bed and the critical area and contributes to the development of microcracks in oil-bearing formations. All this positively influences rheological properties of the fluid [52].

This process analysis has been the focus of many works by a number of authors. However, the mechanism of vibroaction on the bed and the critical area is generally complicated and has not been well analyzed. There is no clear understanding of the factors determining the efficiency of the process [53].

Let us analyze a simplified model and different cases of pulsations of the well flow rate, which are created by the generator (Figure 9.1). It is assumed that the permeability of the formation is dependent on pressure. Then the pressure distribution in the formation is described by the piezoconductivity equation

$$\frac{\partial P}{\partial t} = \frac{1}{r}\frac{\partial}{\partial r}\left[\chi(P)r\frac{\partial P}{\partial r}\right] \quad (9.1)$$

The initial and boundary conditions are

$$P_g\big|_{t=0} = P_b \quad (9.2)$$

$$P\big|_{r=R_{ex}} = P_{ex} \quad (9.3)$$

$$2\pi r h \frac{k}{\mu}\frac{\partial P}{\partial r}\bigg|_{r=r_w} = Q(t) \quad (9.4)$$

The Kirchhoff transformation was used to solve the nonlinear boundary-value problem (9.1):

$$\theta = \frac{1}{\chi_0}\int_0^P \chi(P)\mathrm{d}P \quad (9.5)$$

From Equation (9.1), and accounting for Equation (9.4),

$$\frac{\partial \theta}{\partial t} = \chi_0 \frac{1}{r}\frac{\partial}{\partial r}\left[r\frac{\partial \theta}{\partial r}\right] \quad (9.6)$$

Nanocolloids for Petroleum Engineering: Fundamentals and Practices, First Edition.
Baghir A. Suleimanov, Elchin F. Veliyev, and Vladimir Vishnyakov.
© 2022 John Wiley & Sons Ltd. Published 2022 by John Wiley & Sons Ltd.

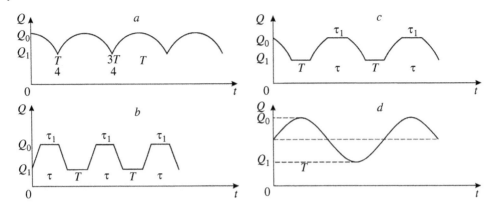

Figure 9.1 Different forms of elastic waves.

9.1 Exact Solution

Based on experimental investigations, it has been shown in reference [52] that near (above) the saturation pressure, the formation of surface gas bubbles leads to a slip effect and this manifests itself in permeability changes. The effective permeability in this case can be determined from the equation

$$k(P) = k_0 \left[1 + \frac{4b(P)}{R}\right] \tag{9.7}$$

The filtration process above and near the saturation pressure is considered (both without allowance for the slip effect and with allowance for it). Linearizing the results obtained in reference [52], for the piezoconductivity it is possible to obtain

$$\chi = \frac{k_0}{\mu\beta^*}\left[1 + \frac{4}{R}a(P - P_w)\right], \quad P_w \leq P \leq P_m; \quad \chi = \frac{k_0}{\mu\beta^*}\left[1 + \frac{1}{R}a(P_s - P)\right], \quad P_m \leq P \leq P_s \tag{9.8}$$

From Equation (9.5), and accounting for Equation (9.8),

$$\theta = P + \frac{2a}{R}\left[(P - P_w)^{2|} - P_w^2\right], \quad P_w \leq P \leq P_m; \quad \theta = P + \frac{2a}{R}\left[P_s^2 - (P_s - P)^2\right], \quad P_m \leq P \leq P_s \tag{9.9}$$

The boundary condition (9.4), and accounting for Equations (9.7) and (9.9), will take the form

$$\frac{k(P)}{\mu} 2\pi hr \frac{\partial P}{\partial r}\bigg|_{r=r_w} = \frac{k_0}{\mu} 2\pi hr \frac{\partial \theta}{\partial r}\bigg|_{r=r_w} \tag{9.10}$$

Equation (9.5), and accounting for the initial Equation (9.2) and boundary conditions (9.3) and (9.4), can be solved using the Duhamel integral [54]. Accounting for Equation (9.10), this was transformed to

$$\Delta\theta(r;t) = Q(0)\Delta\theta_1(r;t) + \int_0^t \dot{Q}(\tau)\Delta\theta_1(r;t-\tau)d\tau \tag{9.11}$$

where $\Delta\theta_1$ is the solution of Equation (9.5) with initial (9.2) and boundary conditions (9.3) and (9.10) for $Q = Q(0) =$ constant.

The solution of Equation (9.5), and taking into account Equations (9.2), (9.3), (9.10), and $Q = Q(0) =$ constant, has the form

$$\Delta\theta_1(r;t) = \frac{Q(0)\mu}{2\pi b k_0}\left[\ln\frac{R_{ex}}{r} + \pi\sum_{\nu=1}^{\infty}\frac{1}{x_\nu}\frac{J_0\left(x_\nu\frac{R_{ex}}{r_w}\right)J_1(x_\nu)U_\nu\left(x_\nu\frac{r}{r_w}\right)}{J_1^2(x_\nu) - J_0^2\left(x_\nu\frac{R_{ex}}{r_w}\right)}\exp\left(-x_\nu^2\frac{\chi_0 t}{r_w^2}\right)\right]$$

(9.12)

where x_ν values are the roots of the transcendental equation

$$J_0\left(x_\nu\frac{R_{ex}}{r_w}\right)Y_1(x_\nu) - J_1(x_\nu)Y_0\left(x_\nu\frac{R_{ex}}{r_w}\right) = 0;$$

$$U_\nu\left(x_\nu\frac{r}{r_w}\right) = J_0\left(x_\nu\frac{r}{r_w}\right)Y_0\left(x_\nu\frac{R_{ex}}{r_w}\right) - J_0\left(x_\nu\frac{R_{ex}}{r_w}\right)Y_0\left(x_\nu\frac{r}{r_w}\right)$$

(9.13)

Integration of Equation (9.11), and accounting for Equation (9.12), at different $Q(t)$ values enables us to find the pressure distribution $P(r, t)$ over the formation bed. Using the elastic wave generator we can create, on the formation bottom, different forms of elastic waves (Figure 9.1a–d) whose equations will respectively be represented as

$$Q = Q_1 + (Q_0 - Q_1)\left|\cos\frac{2\pi}{T}t\right|$$

(9.14)

$$Q = \left\{\frac{2(Q_0 - Q_1)}{\tau - \tau_1}t\left[\eta(t) - \eta\left(t - \frac{\tau - \tau_1}{2}\right)\right] + (Q_0 - Q_1)\left[\eta\left(t - \frac{\tau - \tau_1}{2}\right) - \eta\left(t - \frac{\tau + \tau_1}{2}\right)\right]\right.$$
$$\left. + \left[Q_0 - Q_1 - \frac{2(Q_0 - Q_1)}{\tau - \tau_1}(\tau - t)\right]\left[\eta\left(t - \frac{\tau + \tau_1}{2}\right) - \eta(t - \tau)\right]\right\}$$
$$\times\{\eta[t - (\nu - 1)(T + \tau)] - \eta[t - \nu\tau + (\nu - 1)T]\} + Q_1,$$

(9.15)

$$Q = \left\{(Q_0 - Q_1)\sin\left(\frac{\pi}{\tau - \tau_1}t\right)\left[\eta(t) - \eta\left(t - \frac{\tau - \tau_1}{2}\right)\right] + (Q_0 - Q_1)\left[\eta\left(t - \frac{\tau - \tau_1}{2}\right) - \eta\left(t - \frac{\tau + \tau_1}{2}\right)\right]\right.$$
$$\left. + (Q_0 - Q_1)\cos\left[\frac{\pi}{\tau - \tau_1}\left(t - \frac{\tau + \tau_1}{2}\right)\right]\left[\eta\left(t - \frac{\tau + \tau_1}{2}\right) - \eta(t - \tau)\right]\right\}$$
$$\times\{\eta[t - (\nu - 1)(T + \tau)] - \eta[t - \nu\tau + (\nu - 1)T]\} + Q_1,$$

(9.16)

$$Q = Q_1 + \frac{Q_0 - Q_1}{2}\cos\left(\frac{2\pi}{T}t\right)$$

(9.17)

Then, from Equation (9.11) and accounting for Equation (9.12) and the action function (9.17), which occurs in reality in most cases, the following was obtained

$$\Delta\theta = \frac{(Q_0 + Q_1)\mu}{4\pi h k_0}\left[\ln\frac{R_{ex}}{r} + \pi\sum_{\nu=1}^{\infty}\frac{1}{x_\nu}\frac{J_0\left(x_\nu\frac{R_{ex}}{r_w}\right)J_1(x_\nu)U_\nu\left(x_\nu\frac{r}{r_w}\right)}{J_1^2(x_\nu) - J_0^2\left(x_\nu\frac{R_{ex}}{r_w}\right)}\exp\left(-x_\nu^2\frac{\chi_0 t}{r_w^2}\right)\right]$$

$$+ \frac{(Q_0 - Q_1)\mu}{4\pi h k_0}\ln\frac{R_{ex}}{r}\left(\cos\frac{2\pi}{T}t - 1\right) - \frac{\pi}{T}(Q_0 - Q_1)\frac{\mu}{2hk_0}\sum_{\nu=1}^{\infty}\frac{1}{x_\nu}\frac{J_0\left(x_\nu\frac{R_{ex}}{r_w}\right)J_1(x_\nu)U_\nu\left(x_\nu\frac{r}{r_w}\right)}{J_1^2(x_\nu) - J_0^2\left(x_\nu\frac{R_{ex}}{r_w}\right)}$$

$$\times \left(\frac{\frac{2\pi}{T}\exp\left(-x_\nu^2\frac{\chi_0 t}{r_w^2}\right)}{\frac{x_\nu^4\chi_0^2}{r_w^4} + \frac{4\pi^2}{T^2}} + \frac{\frac{x_\nu^2\chi_0}{r_w^2}\sin\frac{2\pi}{T}t - \frac{2\pi}{T}\cos\frac{2\pi}{T}t}{\frac{x_\nu^4\chi_0^2}{r_w^4} + \frac{4\pi^2}{T^2}}\right) \quad (9.18)$$

If $r_w \leq 0.02 R_{ex}$, which is virtually always observed in this case, formula (9.18) is simplified and takes the form [55]

$$\Delta\theta = \frac{Q_0 - Q_1}{2}\frac{\mu}{2\pi h k_0}\ln\left(\frac{R_{ex}}{r}\right)\left(\cos\frac{2\pi}{T}t - 1\right) + \frac{\pi}{T}(Q_0 - Q_1)\frac{\mu}{hk_0}\left(\frac{r_w}{R_{ex}}\right)^2$$

$$\times \sum_{\nu=1}^{\infty}\frac{J_0\left(x_\nu\frac{r}{r_w}\right)}{x_\nu^2 J_1^2\left(x_\nu\frac{R_{ex}}{r_w}\right)}\left(\frac{\frac{2\pi}{T}\exp\left(-x_\nu^2\frac{\chi_0 t}{r_w^2}\right)}{\frac{x_\nu^4\chi_0^2}{r_w^4} + \frac{4\pi^2}{T^2}} + \frac{\frac{x_\nu^2\chi_0}{r_w^2}\sin\frac{2\pi}{T}t - \frac{2\pi}{T}\cos\frac{2\pi}{T}t}{\frac{x_\nu^4\chi_0^2}{r_w^4} + \frac{4\pi^2}{T^2}}\right) \quad (9.19)$$

$$+ \frac{(Q_0 + Q_1)\mu}{4\pi h k_0}\left[\ln\frac{R_{ex}}{r} - 2\left(\frac{r_w}{R_{ex}}\right)^2\sum_{\nu=1}^{\infty}\frac{J_0(x_\nu)}{x_\nu^2 J_1^2\left(x_\nu\frac{R_{ex}}{r_w}\right)}\exp\left(-x_\nu^8\frac{\chi_0 t}{r_w^2}\right)\right]$$

where x_ν values are the roots of the transcendental equation

$$J_0\left(x_\nu\frac{R_{ex}}{r_w}\right) = 0, \quad \Delta\theta = \theta_{ex} - \theta \quad (9.20)$$

The value of θ_{ex} is determined from Equation (9.9) and has the form

$$\theta_{ex} = P_{ex} + \frac{2a}{R}\left[(P_{ex} - P_w)^{2|} - P_w^2\right], \quad P_w \leq P \leq P_m; \quad \theta_{ex} = P_{ex} + \frac{a}{2R}\left[P_s^2 - (P_s - P_{ex})^2\right],$$
$$P_m \leq P \leq P_s$$

Then for the pressure field in the formation bed, with allowance for the slip effect, we obtain, from Equation (9.9),

$$P = \frac{-\left(1 - \frac{4a}{R}P_w\right) + \sqrt{\left(1 - \frac{4a}{R}P_w\right)^2 + \frac{8a}{R}\theta}}{\frac{4a}{R}}, \quad P_w \leq P \leq P_m \quad (9.21)$$

$$P = \frac{1 + \frac{a}{R}P_s - \sqrt{\left(1 + \frac{a}{R}P_s\right)^2 - \frac{2a}{R}\theta}}{\frac{a}{R}}, \quad P_m \leq P \leq P_s \qquad (9.22)$$

where $\theta = \theta_{ex} - \Delta\theta$, $\Delta\theta$ is determined from the formula (9.18) or (9.19).

When the slip of the fluid is absent, from Equation (9.1), and accounting for Equations (9.2) to (9.4) and (9.19), the pressure field in the formation bed can be expressed as follows:

$$P = P_{ex} - \frac{(Q_0 - Q_1)\mu}{4\pi h k_0} \ln\left(\frac{R_{ex}}{r}\right)\left(\cos\frac{2\pi}{T}t - 1\right) - \frac{\pi}{T}\frac{(Q_0 - Q_1)\mu}{h k_0}\left(\frac{r_w}{R_{ex}}\right)^2$$

$$\times \sum_{\nu=1}^{\infty} \frac{J_0\left(x_\nu \frac{r}{r_w}\right)}{x_\nu^2 J_1^2\left(x_\nu \frac{R_{ex}}{r_w}\right)} \left(\frac{\frac{2\pi}{T}\exp\left(-x_\nu^2 \frac{\chi_0 t}{r_w^2}\right)}{\frac{x_\nu^4 \chi_0^2}{r_w^4} + \frac{4\pi^2}{T^2}} + \frac{\frac{x_\nu^2 \chi_0}{r_w^2}\sin\frac{2\pi}{T}t - \frac{2\pi}{T}\cos\frac{2\pi}{T}t}{\frac{x_\nu^4 \chi_0^2}{r_w^4} + \frac{4\pi^2}{T^2}}\right) \qquad (9.23)$$

$$- \frac{Q_0 + Q_1}{2}\frac{\mu}{2\pi h k_0}\left[\ln\frac{R_{ex}}{r} - 2\left(\frac{r_w}{R_{ex}}\right)^2 \sum_{\nu=1}^{\infty} \frac{J_0(x_\nu)}{x_\nu^2 J_1^2\left(x_\nu \frac{R_{ex}}{r_w}\right)} \exp\left(-x_\nu^2 \frac{\chi_0 t}{r_w^2}\right)\right]$$

9.2 Approximate Solution

With the aim to simplify the resulting formulas (9.18), (9.19), and (9.23), it was found that approximate solution of Equation (9.5) can be produced which allows sufficient accuracy for engineering calculations. Averaging $\partial\theta/\partial t$ over r produces

$$\varphi(t) = \frac{2}{R_{ex}^2 - r_w^2} \int_0^{R_{ex}} \frac{\partial\theta}{\partial t} r \, dt \qquad (9.24)$$

Substituting Equation (9.24) into (9.5), integrating the resulting expression, and accounting for Equations (9.2) to (9.4) and (9.10), we can get

$$\theta = \theta_{ex} + \frac{\varphi(t)}{4\chi_0}(r^2 - R_{ex}^2) - \frac{\varphi(t)}{2\chi_0} r_w^2 \ln\left(\frac{r}{R_{ex}}\right) + \frac{\mu Q(t)}{2\pi h k_0} \ln\frac{r}{R_{ex}} \qquad (9.25)$$

Then from Equation (9.24), accounting for Equation (9.25), and disregarding r_w^2 in the next step,

$$\dot\varphi + \frac{8\chi_0}{R_{ex}^2}\varphi = -\frac{2\dot Q(t)}{\pi \beta^* h R_{ex}^2} \qquad (9.26)$$

The pressure distribution over the formation as a function of r can be found by integrating Equation (9.26). As an example, one can take the function of action on the critical area and the formation bed of Equation (9.17). Then, from Equation (9.26), accounting for Equations (9.17) and (9.25), and the initial condition

$$\theta|_{t=0} = \theta_{ex} + \frac{\mu(Q_1 + Q_0)}{4\pi h k_0} \ln \frac{r}{R_{ex}}$$

it is possible to get

$$\theta = \frac{r^2 - R_{ex}^2}{4\chi_0} \frac{Q_0 - Q_1}{\pi \beta^* h R_{ex}^2} \left\{ -\cos\omega t + \frac{1}{R_{ex}^2} \frac{8\chi_0}{\omega^2 + \left(\frac{8\chi_0}{R_{ex}^2}\right)^2} \left(\frac{8\chi_0}{R_{ex}^2} \cos\omega t + \omega \sin\omega t\right) \right.$$

$$\left. + \left[1 - \frac{\left(\frac{8\chi_0}{R_{ex}^2}\right)^2}{\omega^2 + \left(\frac{8\chi_0}{R_{ex}^2}\right)^2}\right] \exp\left(-\frac{8\chi_0 t}{R_{ex}^2}\right) \right\} + \frac{\mu}{2\pi h k_0} \ln\frac{r}{R_{ex}} \left(Q_1 + \frac{Q_0 - Q_1}{2} \cos\omega t\right) + \theta_{ex}$$

(9.27)

where $\omega = 2\pi/T$.

Next, the pressure distribution over the formation bed with allowance for the fluid's slip is determined from formulas (9.21) and (9.22) for the function θ determined by Equation (9.27). When slip is absent $(P > P_s)$, the pressure distribution over the formation bed is also determined from Equation (9.27) and has the form

$$P = P_{ex} + \frac{r^2 - R_{ex}^2}{4\chi_0} \frac{Q_0 - Q_1}{\pi \beta^* h R_{ex}^2} \left\{ -\cos\omega t + \frac{1}{R_{ex}^2} \frac{8\chi_0}{\omega^2 + \left(\frac{8\chi_0}{R_{ex}^2}\right)^2} \left(\frac{8\chi_0}{R_{ex}^2} \cos\omega t + \omega \sin\omega t\right) \right.$$

$$\left. + \left[1 - \frac{\left(\frac{8\chi_0}{R_{ex}^2}\right)^2}{\omega^2 + \left(\frac{8\chi_0}{R_{ex}^2}\right)^2}\right] \exp\left(-\frac{8\chi_0 t}{R_{ex}^2}\right) \right\} + \frac{\mu}{2\pi h k_0} \ln\frac{r}{R_{ex}} \left(Q_1 + \frac{Q_0 - Q_1}{2} \cos\omega t\right)$$

(9.28)

The bottom-hole pressure can be determined from Equations (9.21) and (9.22) and accounting for Equation (9.27) and from Equation (9.28) for $r = r_w$.

9.3 Concluding Remarks

As is seen from Equations (9.19), (9.23), and (9.28), the waves created by the generator are attenuated as they penetrate deeper into the formation bed. The attenuation of the waves is strongly dependent on the frequency ω. As the frequency of the waves decreases, the depth of their penetration into the bed increases, which is of great practical importance. The propagation of elastic waves with allowance for the slip is nearly of the same character as that without allowance for it. Furthermore, to increase the depth of penetration of the waves, their frequency must be selected to be a multiple of $8\chi_0/R_{ex}^2$.

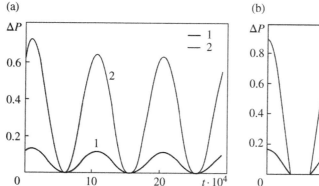
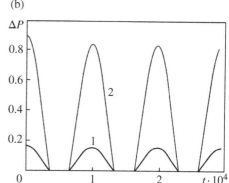

Figure 9.2 Dynamics of the pressure difference ΔP_g for P_{ex} = 200 m: (a) according to the exact solution; (b) according to the approximate solution; (1) with allowance for the slip effect; and (2) without allowance for it. The units are t, s and ΔP, MPa.

Figure 9.3 Depth of penetration of elastic waves vs. vibrational velocity $(A\omega) \times 10^{-4}$ m³ s⁻¹: (1) exact solution; (2) approximate solution. The units are $(A\omega)$, m³ s⁻² and r, m.

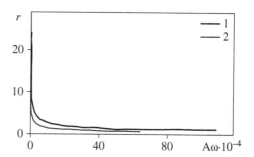

Figures 9.2 and 9.3 give results of numerical calculations, according to the exact and approximate solutions, of the depth of penetration of elastic waves for different $A\omega$ and of the dynamics of the pressure difference for the following values of the system's parameters: $P_{ex} = 10^7$ Pa, $\mu = 1$ mPas, $h = 10$ m, $r_w = 0.15$ m, $\chi_0 = 0.17$ m²s⁻², $k_0 = 0.5 \times 10^{-12}$ m², $a = 10^{12}$ m/Pa, $R = 4.47 \times 10^{-6}$ m, $\beta^* = 2.94 \times 10^{-9}$ 1/Pa, $R_{ex} = 200$ m, $P_3 = 8.5 \times 10^6$ Pa, and $P_w = 5 \times 10^6$ Pa.

From Figure 9.2, it is clear that, all other things being equal, the pressure difference in the process of vibrowave action with the slip effect is always lower than that without it, which is of great practical importance. Indeed, this enables one to maintain a high drainage for relatively low pressure differences [52, 56].

As the frequency of elastic waves decreases, the depth of their penetration into the bed increases (see Figure 9.3). As the product $A\omega$ approaches 10^{-3} m³/s², the depth of penetration of the elastic waves sharply decreases. After this, the penetration depth changes very little and is weakly dependent on $A\omega$. The largest effect from the action is obtained in the interval $0 \leq A\omega \leq 10^{-3}$ m³s⁻².

A comparison of the results of calculations carried out according to the exact and approximate solutions has shown that the approximate solution yields values of the pressure differences (Figure 9.2) and the depth of penetration of elastic waves that are overestimated by approximately 20%. This emboldens one to use the approximate solution for practical engineering calculations.

Nomenclature

A	vibration amplitude, $A = \dfrac{Q_0 + Q_1}{2}$
a	constant coefficient
b	slip coefficient
h	bed thickness, m
J_0 and J_1 and Y_0 and Y_1	Bessel functions of the first and second kind and of zero and first order
$k(p)$	effective permeability, m²
n	constant quantity
P	pressure at any point of the bed, MPa
P_g	pressure on the bed gallery, MPa
P_m	pressure at which the permeability coefficient takes its maximum value, MPa
P_b	bottom-hole pressure, MPa
P_{ex}	pressure at the external boundary of the bed, MPa
P_w	saturation pressure, MPa
P_s	pressure at which the action of the slip effect ceases, MPa
$Q(t)$	flow rate, m³ s⁻¹
Q_0	upper limiting value of the well flow rate, m³ s⁻¹ and derivative of $Q(t)$
Q_1	lower limiting value of the well flow rate, m³ s⁻¹
R	average radius of the pore channel or particle radius, m
R_{ex}	radius of the external boundary of the bed, m
r	coordinate, m
r_w	radius of the well, m
T	period of vibration, s
t	time, s
U_i	suspension velocity of particles
$v_s, v_{ns},$	sedimentation velocity with slippage and without it
β^*	compressibility coefficient, 1 Pa⁻¹
χ	piezoconductivity coefficient, m³ s⁻¹
χ_0	initial value of the piezoconductivity coefficient, m³ s⁻¹

Nanocolloids for Petroleum Engineering: Fundamentals and Practices, First Edition.
Baghir A. Suleimanov, Elchin F. Veliyev, and Vladimir Vishnyakov.
© 2022 John Wiley & Sons Ltd. Published 2022 by John Wiley & Sons Ltd.

η	Heaviside function
$\phi(t)$	unknown function
$\dot{\phi}$	derivative of $\phi(t)$
μ	coefficient of viscosity of the fluid, mPa s
$\nu = 1, 2, 3, ..., n$	natural numbers
θ	Kirchhoff function
τ and τ_1	times characterizing the rate of pulse action, s
ω	frequency of vibrations of elastic waves, s^{-1}

References

1 Thakar, V., Nambiar, S., Shah, M., and Sircar, A. (2018). A model on dual string drilling: on the road to deep waters. *Modeling Earth Systems and Environment* 4 (2): 673–684.
2 Fakoya, M.F. and Shah, S.N. (2017). Emergence of nanotechnology in the oil and gas industry: emphasis on the application of silica nanoparticles. *Petroleum* 3 (4): 391–405.
3 Ozyildirim, C. and Zegetosky, C. (2010). Exploratory investigation of nanomaterials to improve strength and permeability of concrete. *Transportation Research Record* 2142 (1): 1–8.
4 Jalal, M., Mansouri, E., Sharifipour, M., and Pouladkhan, A.R. (2012). Mechanical, rheological, durability and microstructural properties of high performance self-compacting concrete containing SiO_2 micro and nanoparticles. *Materials & Design* 34: 389–400.
5 Yang, X., Shang, Z., Liu, H. et al. (2017). Environmental-friendly salt water mud with nano-SiO_2 in horizontal drilling for shale gas. *Journal of Petroleum Science and Engineering* 156: 408–418.
6 Bahadori, H. and Hosseini, P. (2012). Reduction of cement consumption by the aid of silica nano-particles (investigation on concrete properties). *Journal of Civil Engineering and Management* 18 (3): 416–425.
7 Senff, L., Hotza, D., Lucas, S. et al. (2012). Effect of nano-SiO_2 and nano-TiO_2 addition on the rheological behavior and the hardened properties of cement mortars. *Materials Science and Engineering A* 532: 354–361.
8 Guefrech, A., Mounanga, P., and Khelidj, A. (2011). Experimental study of the effect of addition of nano-silica on the behaviour of cement mortars Mounir. *Procedia Engineering* 10: 900–905.
9 Gaitero, J.J., Campillo, I., Mondal, P., and Shah, S.P. (2010). Small changes can make a great difference. *Transportation Research Record* 2141 (1): 1–5.
10 Qian, S.Z., Zhou, J., and Schlangen, E. (2010). Influence of curing condition and precracking time on the self-healing behavior of engineered cementitious composites. *Cement and Concrete Composites* 32 (9): 686–693.
11 Santra, A.K., Boul, P., and Pang, X. Influence of nanomaterials in oilwell cement hydration and mechanical properties. *SPE International Oilfield Nanotechnology Conference and Exhibition*. Society of Petroleum Engineers (June 2012).
12 Rahimirad, M., Dehghani, B.J. Properties of oil well cement reinforced by carbon nanotubes. *SPE International Oilfield Nanotechnology Conference and Exhibition*. Society of Petroleum Engineers (June 2012).

Nanocolloids for Petroleum Engineering: Fundamentals and Practices, First Edition.
Baghir A. Suleimanov, Elchin F. Veliyev, and Vladimir Vishnyakov.
© 2022 John Wiley & Sons Ltd. Published 2022 by John Wiley & Sons Ltd.

13 Maserati G., Daturi E., Del Gaudio L. et al. Nano-emulsions as cement spacer improve the cleaning of casing bore during cementing operations. In *SPE Annual Technical Conference and Exhibition*. Society of Petroleum Engineers (September 2010).
14 Al-Saba M.T., Al Fadhli A., Marafi A. et al. Application of nanoparticles in improving rheological properties of water based drilling fluids. *SPE Kingdom of Saudi Arabia Annual Technical Symposium and Exhibition*. Society of Petroleum Engineers (April 2018).
15 Mahmoud, O., Nasr-El-Din, H.A., Vryzas, Z., and Kelessidis, V. (2018). Effect of ferric oxide nanoparticles on the properties of filter cake formed by calcium bentonite-based drilling muds. *SPE Drilling & Completion* 33 (04): 363–376.
16 Alvi, M.A.A., Belayneh, M., Saasen, A., and Aadnøy, B.S. The effect of micro-sized boron nitride BN and iron trioxide Fe_2O_3 nanoparticles on the properties of laboratory bentonite drilling fluid. In *SPE Norway One Day Seminar*. Society of Petroleum Engineers (April 2018).
17 Minakov, A.V., Zhigarev, V.A., Mikhienkova, E.I. et al. (2018). The effect of nanoparticles additives in the drilling fluid on pressure loss and cutting transport efficiency in the vertical boreholes. *Journal of Petroleum Science and Engineering* 171: 1149–1158.
18 Amanullah, M., Al Arfaj, M.K., Al-Abdullatif, Z.A. Preliminary test results of nano-based drilling fluids for oil and gas field application. *SPE/IADC Drilling Conference and Exhibition*. Society of Petroleum Engineers (March 2011).
19 Amanullah, M. and Al-Tahini A.M. Nano-technology-its significance in smart fluid development for oil and gas field application. *SPE Saudi Arabia Section Technical Symposium*. Society of Petroleum Engineers (May 2009).
20 Hoelscher, K.P., De Stefano, G., Riley, M., and Young, S. Application of nanotechnology in drilling fluids. In *SPE International Oilfield Nanotechnology Conference and Exhibition*. Society of Petroleum Engineers (June 2012).
21 Sayyadnejad, M.A., Ghaffarian, H.R., and Saeidi, M. (2008). Removal of hydrogen sulfide by zinc oxide nanoparticles in drilling fluid. *International Journal of Environmental Science and Technology* 5 (4): 565–569.
22 Bhatia, K.H. and Chacko, L.P. Ni-Fe nanoparticle: An innovative approach for recovery of hydrates. *Brasil Offshore*. Society of Petroleum Engineers (May 2011).
23 Kumar, D., Chishti, S.S., Rai, A., Patwardhan, S.D. Scale inhibition using nano-silica particles. *SPE Middle East Health, Safety, Security, and Environment Conference and Exhibition*. Society of Petroleum Engineers (April 2012).
24 Cui, Z., Yin, L., Wang, Q. et al. (2009). A facile dip-coating process for preparing highly durable superhydrophobic surface with multi-scale structures on paint films. *Journal of Colloid and Interface Science* 337 (2): 531–537.
25 Ye, Q., Zhang, Z.N., Chen, R.S., and Mac, C. (2003). Interaction of nano-SiO_2 with calcium hydroxide crystals at interface between hardened cement paste and aggregate. *Journal of the Chinese Ceramic Society* 31 (3).
26 Lin, D.F., Tsai, M.C., and Air, J. (2006). Study of recycled materials. *The Official Journal of the Air and Waste Management Association* 56 (5): 1140–1146.
27 Tao, J. (2012). Preliminary study on the water permeability and microstructure of concrete incorporating nanoSiO_2. *Cement and Concrete Research* 35 (10): 1943–1947.
28 Ye, Q. (2001). Research on the comparison of pozzolanic activity between nano-SiO_2 and silica fume. *Concrete Journal* 3 (1): 19–22.

29 Chen, R.S. and Ye, Q. (2001). Research on the comparison of properties of hardend cement paste between nano-SiO_2 and silica fume added concrete. *Concrete Journal* 3 (1): 19–22.
30 Li, H., Xiao, H., and Ou, J. (2004). Nono-particled in concrete. *Cement and Concrete Research* 34 (3): 435–438.
31 Li, H. and Zhang, M.J.O. (2006). Abrasion resistance of concrete. *Wear* 260 (3): 1262–1266.
32 Nazari, A. and Riahi, S. (2010). The effect of TiO_2 nanoparticles on water permeability and thermaland mechanical properties of high strength self-compacting concrete. *Materials Science and Engineering A* 52 (8): 756–763.
33 Nazari, A. and Riahi, S. (2010). Improvement compressive strength of concrete in different curing media by Al_2O_3 nanoparticles. *Materials Science and Engineering A* 52 (8): 1183–1191.
34 Nazari, A. and Riahi, S. (2010). The role of SiO_2 nanoparticles and ground granulated blast furnace slag admixtures on physical, thermal and mechanical properties of self compacting concrete. *Materials Science and Engineering A* 52 (8): 2149–2157.
35 Nazari, A. and Riahi, S. (2010). The effects of Cr_2O_3 nanoparticles on strength assessments and water permeability of concrete in different curing media. *Materials Science and Engineering A* 528 (3): 1173–1182.
36 Baolin, Z.H.U., Xin, H., and Ye, G.U.O. (2011). Influence of cement particle size distribution on strength of hardened cement paste. *Key Engineering Materials* 477: 118–124.
37 Bentz, D.P. (2010). Blending different fineness cements to engineer the properties of cement-based materials. *Magazine of Concrete Research* 62 (5): 327–338.
38 Frigione, G. and Marra, S. (1976). Relationship between particle size distribution and compressive strength in Portland cement. *Cement and Concrete Research* 6 (1): 113–127.
39 Celik, I.B. (2009). The effects of particle size distribution and surface area upon cement strength development. *Powder Technology* 6 (3): 272–276.
40 Bentz, D.P., Garboczi, E.J., Haecker, C.J., and Jensen, O.M. (1999). Effects of cement particle size distribution on performance properties of Portland cement-based materials. *Cement and Concrete Research* 28 (10): 1663–1671.
41 Partcile size analysis of cement using the technique of laser diffraction. (2014). UK, Malvern Instruments.
42 Ginebra, M.P., Driessens, F.C.M., and Planell, J.A. (2004). Effect of the particle size on the micro and nanostructural features of a calcium phosphate cement: A kinetic analysis. *Biomaterials* 25 (17): 3453–3462.
43 Sanchez, F. and Sobolev, K. (2010). Nanotechnology in concrete – A review. *Construction and Building Materials* 24: 2060–2071.
44 Suleimanov, B.A. and Veliyev, E.F. (2016). The effect of particle size distribution and the nano-sized additives on the quality of annulus isolation in well cementing. *SOCAR Proceedings* 4: 4–10.
45 Lawson, R. (1996). How to perform safer hydraulic workovers in gas wells. *World Oil* 217 (1): 42–45.
46 Suman, G.O. Jr., Ellis, R.C., and Snyder, R.E. (1983). *Sand Control Handbook*, 2e. London: Gulf Publishing Company.
47 Davidson, J.F., Clift, R., and Harrison, D. (1985). *Fluidization*, 2e. London: Academic Press.
48 Happel, J. and Brenner, H. (1983). *Low Reynolds Number Hydrodynamics*. Springer Science +Business Media B.V.

49 Zhou, Z., Xu, Z., Finch, J.A., and Liu, Q. (1996). Effect of gas nuclei on the filtration of fine particles with different surface properties. *Colloids and Surfaces A* 113: 67–77.

50 Boehnke, U.-C., Remmler, T., Motschmann, H. et al. (1999). Partial air wetting on solvophobic surfaces in polar liquids. *Journal of Colloid and Interface Science* 211: 243–251.

51 Suleimanov, B.A. (2006). *Filtration Features of Heterogeneous Systems*. In: *Izhevsk Institute of Computer Science*, vol. 2006. ISBN: 5-93972-592-9.

52 Suleimanov, B.A. (1997). Slip effect during flow of gassed liquid. *Colloid Journal* 59 (6): 749–753.

53 Baishev, E.V., Glivenko, E.V., Gubar, V.A. et al. (2004). Gas-pulse action on the well bottomhole zone. *Fluid Dynamics* 39: 582–588.

54 Slezkin, N.A. (1955). *Dynamics of Viscous Incompressible Fluids*. Moscow: Gostekhizdat.

55 Watson, G.N.A. (1966). *Treatise on the Theory of Bessel Functions*, 2e. Cambridge University Press.

56 Suleimanov, B.A., Abbasov, E.M., and Efendieva, A.O. (2008). Vibrowave action on the bed and the critical area of wells with allowance for the slip effect. *Journal of Engineering Physics and Thermophysics* 81 (2): 380–387.

Part D

Enhanced Oil Recovery

10

An Overview of Nanocolloid Applications for EOR

A substantial number of studies recently reported about perspectives and advantages of nanocolloid applications in petroleum engineering. Conventionally dealing with nano- and microscale phenomena enhanced oil recovery (EOR) methods gained large experience and data compared to other areas of the petroleum industry in this regard. There are a variety of contribution mechanisms of nanocolloids on the oil recovery process (e.g. wettability alteration, interfacial tension reduction, increasing the physical properties of polymer systems, etc.). This chapter will discuss above-mentioned topics in detail and provides a theoretical basis for the nanoscale phenomena impact on colloids that could be used in the EOR.

The majority of studies performed in the last decade have mainly focused on wettability alterations. Meanwhile, the viscosity reduction has also been and is one of the central research topics. Thus Chen et al. showed the successful implementation of nanoparticles as catalysts to produce from oil the components containing aromatic and saturated hydrocarbons. The mechanism behind the presented results was explained by oil composition alteration in the presence of nanoparticles, leading to reduction of asphaltene and resin contents. This leads to the oil light part increment, which then acts as the solvent and decreases the oil viscosity [1]. The nano-dispersion phase impact on oil's asphaltene content was also studied by Nassar et al. The researchers similarly concluded that nanoscale additives demonstrate catalytic and oxidation abilities as well as spontaneous adsorption of asphaltene on nanoparticle surfaces. Ogolo et al. performed a study with a wide range of nanoparticles to investigate their performance on hydrocarbon production [2]. Many oxides and compounds were used: magnesium oxide, aluminum oxide, zinc oxide, zirconium oxide, tin oxide, iron oxide, nickel oxide, hydrophobic silicon oxide, and silicon oxide treated with silane. Authors reported interfacial tension and oil viscosity reduction associated with nanoparticles addition. The dispersion medium fluid importance for observed phenomena was also noted.

The majority of the presented studies were laboratory-based studies. However, reservoir conditions modeling the capability of core flood experiments is a main key for evaluation of nanocolloids effectiveness as an EOR agent. Only experiments at modeled reservoir

---

Nanocolloids for Petroleum Engineering: Fundamentals and Practices, First Edition.
Baghir A. Suleimanov, Elchin F. Veliyev, and Vladimir Vishnyakov.
© 2022 John Wiley & Sons Ltd. Published 2022 by John Wiley & Sons Ltd.

conditions make the studies very relevant and essential for reservoir engineers. In general, these experiments could be divided into two groups:

- Core flooding experiments focused on dispersion phase properties.
- Core flooding experiments focused on dispersion medium properties.

10.1 Core Flooding Experiments Focused on Dispersion Phase Properties

Experiments focused on dispersion phase study concentration and type of nanoscale additives influence on oil recovery. The concentration of nanoparticles is generally predefined while initial screening tests, like interfacial tension (IFT) measurements, rheology, and wettability, are measured. The main goal behind this screening is to reduce the number of time-consuming core flooding studies. It is evident that there is a critical concentration of nanoparticles above which the effect on oil recovery is not observed and even reduced compared to conventional method applications. For instance, Ehtesabi et al. presented a study that shows 31% of oil production increases while using titania (0.01 wt%) compared to low salinity water injection (e.g. 5000 ppm brine). However, a further increase in titania concentration up to 1% leads to decrement in oil recovery with reference to brine injection only [3]. Meanwhile, the researchers mostly prefer to use lower density nanoparticles rather than titania. This is related to the phenomena when high-density nanoparticles make the colloid stabilization quite challenging [4]. For instance, Son et al. used silica (3 wt%) nanoparticles to stabilize a polyvinyl alcohol (PVA) emulsion and achieved a 4% increase in oil recovery [5]. Maghzi et al. showed that addition of silica nanoparticles to injected water leads to a 8.7–26% increment in oil recovery, depending on the nanoparticle concentration [6]. Bayat et al. presented a comparative study of the impact of the addition of metal oxide nanoparticles on oil production at different reservoir conditions. Very high growth in oil recovery was observed with: alumina nanofluid – 52.6% OOIP; titania – 50.9%, OOIP; and silica – 48.7% OOIP. The highest recovery by alumina nanofluids the authors attributed in their obtained results was to the ability of alumina nanofluid to decrease capillary forces during oil displacement [7]. This was also confirmed by the study presented by Alomair et al. when the advantages of titania nanoparticles application compared to silica nanoparticles were not achieved. It was observed that the highest oil recovery is associated with alumina nanofluid (e.g. 4.895%. OOIP) but the mixture of silica and alumina shows even better results (e.g. 23.724%) [8]. Nazari et al. reported an oil recovery increase by a factor of 6 when the base fluid contains silica nanoparticles. In addition, eight different nanoparticle wettability performances were investigated within this study. Calcium carbonate and silica nanoparticles were selected as the best performers for the EOR application. In summary, various nanoparticles have different impacts on the oil production process, depending on the oil composition and reservoir conditions.

10.2 Core Flooding Experiments Focused on Dispersion Medium Properties

Dispersion medium properties is another central topic in EOR-related studies with nanocolloid applications [9]. Suleimanov et al. investigated the effect of alumina nanoparticle addition to surfactant aqueous solution and obtained 22 and 17% increments in oil recovery compared to water and surfactant injection respectively. The authors stated that implementation of nanofluid significantly decrease the IFT and alter the rheological behavior [10]. Meanwhile, addition of hydrophilic and slightly hydrophobic silica nanoparticles also leads to performance improvement by surfactant aqueous solution injection [11]. Sun et al. presented a foam stabilization effect of silica nanoparticles with sodium dodecyl sulfate (SDS) addition. SDS particles improve foam stability and rheology; an increase in differential pressure also leads to an injection profile modification effect [5]. Pei et al. reported that a silica surfactant-stabilized emulsion could act as an effective displacement agent in EOR processes [12].

The mineralization (e.g. water hardness) of connate water as well as injected water plays an important role during nanoparticle application. Obtained data shows that an increase in water mineralization is associated with a higher oil recovery but beyond some critical point it leads to nanoparticle colloid destabilization [13–15].

The performance of injection fluids under reservoir conditions is very critical for polymer injection-based EOR methods due to the sensitivity of polymeric fluids to reservoir mineralization and temperature. Thus, the in-situ stability improvement of polymer-based agents is one of the main challenges for petroleum engineers. Nanoparticle additives could significantly adjust the physical–chemical properties of such systems and would be a good way to resolve instability issues. This topic will be discussed in detail further on in the book.

11

Surfactant-Based Nanofluid

A new type of fluid, usually called a "smart fluid," has become more accessible for the oil and gas industry [16, 17]. The nanofluids are created by the addition of nanoparticles to fluids for intensification and improvement of some properties at low volume concentrations of the dispersing medium. The main feature of nanofluids is that their properties greatly depend on the nanoparticle dimensions that are their concentrations [18]. Nanoparticle suspensions have the following advantages: an increase in sedimentation stability due to the fact that the surface forces easily counterbalance the force of gravity; the thermal, optical, stress–strain, electrical, rheological, and magnetic properties that strongly depend on the size and shape of the nanoparticles. The latter set of properties can be relatively easy to select during production. It is for this reason that nanofluid properties often exceed the properties of conventional fluids [19–21]. The results of many experimental works show how dispersed nanoparticles in an aqueous phase can modify the interfacial properties of the liquid/oil systems if their interfaces are modified by the presence of an ionic surfactant. Mixed particle/surfactant interfacial layers have been characterized and add yet another dimension in the possibility to adjust interfacial tension and wettability.

All results reported below are related to the interfaces of oil in diluted dispersions of non-ferrous metal nanoparticles in an anionic surfactant (sulphanole – alkyl aryl sodium sulphonate) aqueous solution.

11.1 Nanoparticle Influence on Surface Tension in a Surfactant Solution

The presence of nanoparticles changes rheological properties and increases the effect of the surfactant solution on oil recovery processes. Most importantly, it changes the interfacial tension within the surfactant/oil interface [22]. An observed reduction of the interfacial tension is the result of the presence of nanoparticles at the interfacial layers. At low nanoparticle concentrations, the nanoparticles are attached to the liquid surface and, due to the absorption process, decrease surface tension. However, in concentrations larger than 0.4 wt%, the nanoparticles nearly completely remove the surfactant from the bulk aqueous phase and there is no free surfactant available in the bulk. Thus, for nanoparticle concentrations below 0.4 wt%, the interfacial tension of the dispersion is determined by a mixed

---

Nanocolloids for Petroleum Engineering: Fundamentals and Practices, First Edition.
Baghir A. Suleimanov, Elchin F. Veliyev, and Vladimir Vishnyakov.
© 2022 John Wiley & Sons Ltd. Published 2022 by John Wiley & Sons Ltd.

11 Surfactant-Based Nanofluid

Table 11.1 Nanoparticle influence on the change of surface tension in surfactant solution.

Concentration (wt%)		Surface tension (10^{-3} N m^{-1})	
Sulphanole	Nanoparticles	Surfactant solution	Surfactant solution with nanoparticles addition
0.004	0.001	31.4	9.2
0.0078	0.001	18.4	5.5
0.0156	0.001	16.6	3.6
0.0312	0.001	14.7	1.8
0.05	0.001	10.9	1.09

layer composed of attached nanoparticles and surfactant adsorbed at the liquid interface [23].

In the first series of experiments, the interfacial (surface) tension, adsorption, and wettability of prepared formulations were determined. The following methods of measurement were selected:

1) The pendant drop method is based on the Laplace equation, which describes the relationship between the difference in pressure and interfacial tension.
2) The sessile drop method is an optical contact angle method. This method is used to estimate the wetting properties of a localized region on a solid surface.

The results of the wetting angle and the interfacial tensions of the aqueous solution–oil interface were produced by using the drop shape analysis method at 298 K (Table 11.1).

Nanoparticle addition decreases surface tension by 70–79% at very small concentrations of 0.004–0.0078 wt% in sulphanole solution. At sulphanole concentrations above 0.0156 wt%, the mass effect of nanoparticle addition becomes even more pronounced and the surface tension is reduced by 88–90%.

11.2 Nanoparticle Influence on the Surfactant Adsorption Process

The adsorption process of a surface-active agent was studied under static conditions on silica sand specimens with grains of 0.315–0.2 mm diameter, using 10 g of silica sand and 100 ml of surfactant aqueous solution. Two different surfactant concentrations in solution were used of 0.0078%–0.05 wt%, where the nanoparticle concentration was 0.001 wt%. The experiments were conducted at 298 K. The surfactant absorption was measured by the change of interfacial tension. The measurements were performed every 24 hours during 3 days from the time of solution preparation. Values shown in Table 11.2 demonstrate the dynamics of surfactant absorption.

In the solution with just sulphanole, after the first 24 hours some surface-active agent desorption from interfaces was observed, and the concentration was then shown to

Table 11.2 Nanoparticle influence on the surfactant adsorption process.

Concentration (% mass)		Surface tension (10^{-3} N m^{-1})				Adsorption limit (mg g^{-1})
Sulphanole	Nanoparticles	Initial	24 (h)	48 (h)	72 (h)	
0.0078	0	18.4	20	19	19	0.04
0.0078	0.001	5.5	11	11	11	0.58
0.05	0	10.9	12.1	11.8	11.8	0.04
0.05	0.001	1.09	3.2	3	3	0.74

stabilize. In the presence of nanoparticles the sulphanole adsorption process is more stable and surfactant adsorption values went up 14.5 and 18.5 times respectively.

11.3 Nanoparticle Influence on Oil Wettability

Oil wettability in the presence of surface-active agents and nanoparticles was measured. The sulphanole concentration was 0.05 wt% and the nanoparticle concentration was 0.001 wt%. The experiments were performed at 298 K. The results are presented in Table 11.3. As shown in the table, oil wettability remains practically unchanged after the nanofluid addition.

11.4 Nanoparticle Influence on Optical Spectroscopy Results

Absorption spectra were obtained with a Lambda-40 spectrophotometer, manufactured by Perkin Elmer, USA, over the range 190–1100 nm. For optical spectroscopy results we intentionally used a less effective nanofluid composition in order to show clear differences in the absorption process due to nanoparticle presence, even in that case.

To reveal the mechanism of the observed phenomena in the second series of experiments, optical spectroscopy of the solutions was carried out. An aqueous surfactant solution with 0.0078% mass of sulphanole was used for this series of experiments.

Table 11.3 Nanoparticle influence on oil wettability.

	Concentration (% mass)		
Oil	Sulphanole	Nanofluid	Wettability (cos θ)
100	0	0	0.7604
100	0.05	0	0.918
100	0.05	0.001	0.945

The infrared (IR) spectra of sulphanole powder and sulphanole with a nanostructurized component isolated from aqueous solutions are shown in Figure 11.1a, b.

In Figure 11.1a, in the IR sulphanole spectrum absorption bands (AB) at 1166 and 1049 cm^{-1}, S—O bond vibrations and 650 cm^{-1} —C—S bond vibrations were found. The bands at 735 and 695 cm^{-1} are related to substituted aromatic C—H bonds of the benzene ring. For the sulphanole powder with a nanostructurized component, isolated from aqueous solutions, no significant changes of bands belonging to sulphanole were revealed.

Ultraviolet (UV) and visible (VIS) spectra of a surfactant aqueous solution at a sulphanole concentration of 0.0078 wt% and the same solution with a nanostructurized powder with 2.5×10^{-4} wt% concentration are shown in Figure 11.2a, b. It can be seen that the UV/VIS spectra of sulphanole aqueous solution are characterized by absorption bands at 262 and 233 nm. The first band belongs to sulphanole, while the second band is mainly like an admixture of organic structures. Introduction of only 2.5×10^{-4}% mass of nanostructurized powder into sulphanole aqueous solution causes noticeable changes in the absorption spectrum. Bands belonging to the sulphanole aqueous solution practically disappear and instead two intensive bands appear at 223.0 (1.00) and 200 (1.09) nm, the intensity of which is two-fold higher. The experiments showed that an increase of the nanostructurized component in solution from 2.5×10^{-4} up to 5.0×10^{-3}% mass did not materially affect the absorption intensity. For a sulphanole aqueous solution containing 0.01% mass of nanostructurized powder, a noticeable increase of absorption bands was observed at 200 nm (1.25) whereas the intensity of the absorption bands at 223.0 nm hardly changed. The experiments showed that increasing the nanostructurized component in solution from 2.5×10^{-4} up to 5.0×10^{-3} %mass did not materially affect the absorption intensity.

The results described here unambiguously show that the change of initial absorption bands belonging to a sulphanole aqueous solution in the presence of a nanostructurized component is controlled by the interaction of sulphanole anions ($C_{12}H_{25}$ —C_6 H_4 —SO_3) with cations of the nanopowder surface.

To find out the nature of this interaction and the structure of sulphanole anion complexes with surface hydroxyls of the nanostructurized component, first of all it is necessary to pay attention to the following.

During introduction of nanostructurized powder into sulphanole aqueous solution: (i) absorption bands are moved toward the lower frequency (hypsochromic shift) and (ii) absorption band intensity significantly increases (the hyperchromic effect). The mechanism of interaction, adsorption of surfactant molecules on solid surfaces, can be studied by detailed research of the behavior of electronic and oscillating spectra absorption bands of these molecules and the adsorbent. From given UV/VIS and IR-spectra (Figure 11.1) it is clearly shown that surface-active agent adsorption causes noticeable change in the position of absorption bands in IR and absorption electronic spectra in comparison with the surface-active agent spectrum. The IR-spectra wide absorption band with maximal 3350 cm^{-1} (Figure 11.1a, b) is shown against a background of band intensity decrease at 3500–3400 cm^{-1}. Washing the powders by sonification for one hour in water causes some changes in the spectrum. The band intensity is decreased in the region of 2800–3000 cm^{-1}, while the intensity of the hydroxyl residual band remains practically unchanged. Holding specimens at a temperature of 303.15–313.15 K during two weeks did not materially change the

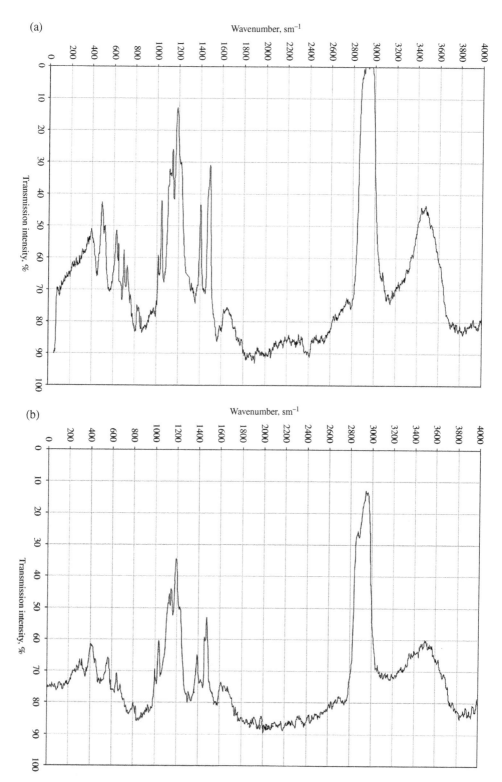

Figure 11.1 IR spectrum of powder separated from aqueous solution: (a) sulphanole; (b) sulphanole with nanopowder addition.

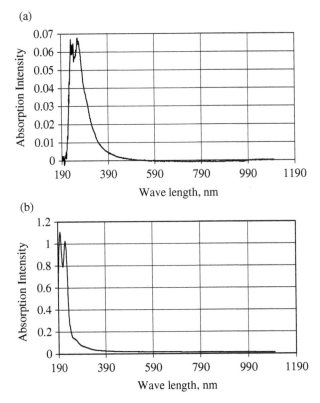

Figure 11.2 Saturation spectrum UV/VIS of aqueous solution: (a) sulphanole; (b) sulphanole with nanopowder addition.

spectrum shape. The experiments conducted show surfactant chemosorption at the initial stage of adsorption and physical adsorption at the subsequent stages.

11.5 Nanoparticle Influence on the Rheological Properties of Nanosuspension

Rheological properties of nanosuspension were determined on a rotational viscosimeter PVS of Brookfield Engineering at 298 K. The PVS is a dynamic coaxial cylinder with a controlled shear rate rheometer. The outer cylinder (sample cup) is driven by speeds from 0.05 to 1000 rpm. The inner cylinder (bob) contains an RTD probe on the surface to provide temperature measurements where the shear stress is being measured. Several cup and bob designs with different geometries are available to suit various applications.

The dependence of shear stress on shear rate for aqueous solutions of anionic surfactant with and without the addition of nanoparticles is shown on Figure 11.3. As may be seen from the figure, nanoparticle addition causes modification of the flow character from Newtonian to non-Newtonian (pseudoplastic), i.e. the obtained nano-fluid is characterized by

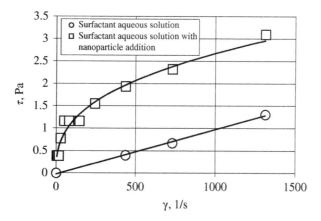

Figure 11.3 Rheological curve of anionic surfactant aqueous solution without addition and with addition of nanoparticles.

shift dilution. The minimal Newtonian viscosity of the surfactant aqueous solution was 0.98 mPa s, but with added nanoparticles it became twice higher and was equal to 2 mPa s. The surfactant aqueous solution was 0.05 wt% sulphanole concentration.

11.6 Nanoparticle Influence on the Processes of Newtonian Oil Displacement in Homogeneous and Heterogeneous Porous Mediums

Tests simulating the processes of Newtonian oil displacement in homogeneous and heterogeneous porous mediums were carried out on the experimental setup shown -in Figure 5.1.
Tests were carried out according to the following plan:

- The column of high pressure was filled with quartz sand of the required fraction, the unit was connected in accordance to Figure 5.1, permeability to the air was determined, and vacuum treatment of the whole unit was carried out at a constant thermostat setting.
- The porous medium was saturated with water that was displaced by oil with a viscosity of 7 mPa s at 298 K up to complete saturation with a simultaneous measurement of the stratum pore volume.
- The required pressures were set on the inlet and outlet to the column and oil was filtered; the oil was then passed through at a constant flow rate. This enabled the permeability of the porous medium to oil to be determined.
- The PVT chamber was filled by displacement fluid (tap water, water with sulphanole addition and nanoparticles), depending on the test purpose.
- Oil was displaced by the displacement fluid at a constant pressure drop and the quantity of displaced oil was measured.

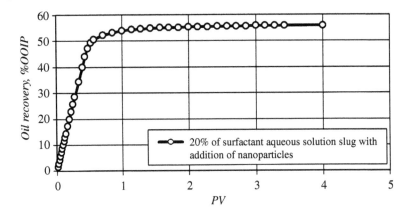

Figure 11.4 Oil recovery curve in a stratified heterogeneous porous medium.

Tests were carried out in the following porous media:

- Homogeneous with permeability 1 D and porosity 26%.
- Stratified and heterogeneous with contacting homogeneous layers; the permeability of the layers was 0.4 and 1.6 D with porosity of 20 and 28% respectively, which was achieved by gradually removing the partition in the process of column filling.

A comparison of the results showed that the mean of the heterogeneous porous medium permeability was chosen to be equal to that of the homogeneous medium. Displacement was carried out at a pressure drop of 0.1 MPa at 298 K.

Three different displacement agents were used for this series of experiments: water, aqueous solution of surfactant (0.05 wt% of sulphanole solution), and nanofluid (0.05 wt% of sulphanole solution and 0.001 wt% of nanoparticles). The results were registered for oil recovery, with %OOIP dependent on the pore volume of the injected fluid and are shown in Figures 11.4 and 11.5 and Table 11.4.

In a homogeneous porous medium at water-free oil recovery, nanofluid increased oil recovery (%OOIP) to 51 and 35% accordingly with the water and surfactant solution. The finite oil recovery (%OOIP) increase was 17% compared to water and 12% compared to surfactant solution.

In a heterogeneous porous medium at water-free oil recovery, the nanofluid as a displacement agent increased oil recovery (%OOIP) by more than 66% compared to water, but the finite oil recovery increase was 17 and 22% compared to the surfactant solution and water respectively.

It should be noted that at the oil recovery by a nanofluid slug (20% from pore volume), water-free and finite oil recovery was the same as analogous parameters at a constant recovery by the nanofluid.

The tests were conducted in the same setup. Oil with a viscosity of 20 mPa s at 298°K was utilized. The content of the oil heavy components was 17%. The porous medium was saturated with formation water that was displaced by the crude oil. When only pure oil was produced at the outlet, the residual water volume was 28–30% of the pore volume. Then crude oil was displaced by the nanofluid. For the purpose of comparison, the same tests

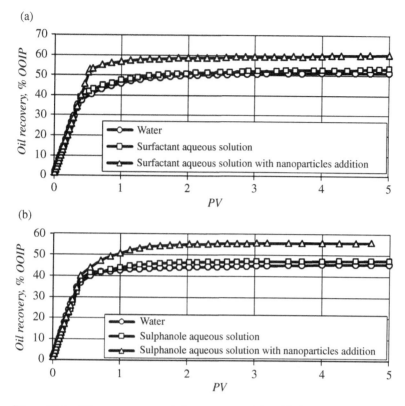

Figure 11.5 Oil recovery curves in homogeneous (a) and in layered heterogeneous (b) porous medium.

Table 11.4 Water-free and finite factors of oil recovery and %OOIP for different displacement agents.

Displacement agent	Oil recovery, %OOIP in homogeneous porous medium		Oil recovery, %OOIP in heterogeneous porous medium	
	Water-free % OOIP	Finite % OOIP	Water-free % OOIP	Finite % OOIP
Water	35	51.3	26.4	45.8
Aqueous solution of surfactant	39	53.4	22.2	47.5
Nanofluid	53	60.3	44	56
Nanofluid slug	–	–	44	56

were conducted using water and an aqueous solution of anionic surfactant as the working agent. The test results are shown in Table 11.5.

It is obvious from Table 11.5 that the production rate of oil displaced by the nanofluid is increased almost 1.5-fold in comparison with the aqueous solution of the anionic

Table 11.5 Oil flow rate.

Working agent	Additives	Concentration	Oil flow rate (cm^3 s^{-1})
Water	Water	100	0.015
Aqueous solution of surfactant	Surfactant	0.05	0.048
Nanofluid	Surfactant	0.05	0.07
	Nanoparticles	0.001	

surface-acting agent and 4.7-fold in comparison with water. This is despite the fact that the minimal nanofluid Newtonian viscosity is only two-fold greater than the viscosity of water and the surfactant aqueous solution. It is obvious that the decrease in interfacial tension on the nanofluid–oil interface and improvement in pore wettability cause an energy reduction of oil in contact with the porous medium surface. Due to the latter fact, the oil flow rate has increased.

11.7 Concluding Remarks

1) Nanoparticles decrease surface tension by 70–79% with 0.004–0.0078 wt% concentration in sulphanole solution, but if sulphanole concentration is more than 0.0156 wt% then the surface tention is reduced by 88–90%.
2) In the presence of nanoparticles the sulphanole adsorption process is more stable and surfactant adsorption values exceed those observed in a surfactant solution by 14.5–18.5 times.
3) Nanoparticles added to the sulphanole solution do not practically change oil wettability.
4) Nanoparticles are attached to the liquid surface and decrease surface tension. The results of spectrographic measurements show the change of initial light absorption bands belonging to a sulphanole aqueous solution in the presence of a nanostructurized component. The process is controlled by the interaction of sulphanole anions ($C_{12}H_{25}$–C_6H_4–SO_3) with cations of the nanopowder surface.
5) Nanoparticles added to the surface-active agent solution cause flow character modification from Newtonian to non-Newtonian (pseudoplastic), i.e. the nanofluid is characterized by the observed shift dilution. The minimal Newtonian viscosity of the surfactant aqueous solution with added nanoparticles is twice as high as without the nanoparticles and is equal to 2 mPa s.
6) In a homogeneous pore medium water-free oil recovery the increase was 51 and 35% and finite was 17 and 12% according to the water and surfactant solution. In a heterogeneous pore medium the water-free oil recovery increase was 66% compared to water and finite was 22 and 17% according to the water and surfactant solution. It should be noted that at oil recovery by a nanofluid slug (20% from pore volume), water-free and finite oil recovery was the same as analogous parameters at constant recovery by a nanofluid.
7) Application of the nanosuspension developed here permitted a significant increase in the efficiency of the oil displacement flow rate. The production rate of oil displaced by the nanofluid increased almost 1.5-fold in comparison with the aqueous solution of an anionic surface-acting agent and 4.7-fold in comparison with water.

12

Nanofluids for Deep Fluid Diversion

The recovery factor of all enhanced recovery operations in heterogeneous reservoirs with high permeability streaks can be increased by the application of gelled polymer systems. In contrast to near wellbore treatment applications, polymer gels are generally applied in injector wells through in-depth fluid diversion (IFD) to increase reservoir-wide sweep efficiency. Novel gel systems allow multichoice range solutions from weak gels to sequential injection for in situ gels, colloidal dispersion gels (CDGs), preformed gels, and microgels. The objective of an IFD process is to modify the reservoir inflow profile for significant reduction of thief zones with effective permeability to prevent the water uptake.

The number of studies considering the problem of unswept zone activation has significantly increased in the last 25 years. The field trials of CDG application for in-situ fluid diversion have been presented by different authors [24, 25]. The CDG as compared to a bulk gel do not have continuous intermolecular gel networks. The network essentially does not develop due to the low polymer concentration in the composition. The system is a microscale separate inhomogeneous entity with uniformly dispersed gel/colloid globules, formed by a cross-linking reaction. This arrangement reduces the motional degrees of freedom as compared to a pure polymer.

Another branch of polymer systems, used for IFD operations, is that of the preformed particle gels (PPGs). These cross-linked, three-dimensional networks of polymer chain materials can absorb large quantities of water [25]. The PPGs are the newest materials for gel treatment methods, which are environmentally friendly and have high water absorption properties.

In general, gels should easily be able to propagate into high permeability zones. This high permeability is linked to fine gel particle migration through the porous media. Some part of the particle passes through the pore throats and propagate forward, while the other part is trapped and blocks the flow. These blocks divert the subsequent injected fluid into low permeability zones and displace residual saturated oil. Overall, this increases the reservoir oil recovery.

12.1 Pre-formed Particle Nanogels

Polymer gel injection can be classified as either bulk gel injection (surface produced) or sequential injection (in-situ). Each of these two injection approaches has key disadvantages and benefits. Bulk injection involves the injection of a homogeneous gel solution, formulated at the surface before the injection process takes place. This homogeneous gel solution, injected into the formation, quickly becomes a strong gel in-situ and predominantly affects the near-wellbore region. Bulk injection can result in weaker gels as a result of shear degradation. Sequential injection involves sequential injections of polymer and crosslinker. It enables deep placement of gels, as there is not much chance for crosslinking until both the polymer and crosslinker are injected into the formation. Only then will the polymer crosslinking take place, yielding a strong gel capable of varying a zone's permeability. The disadvantage of sequential injection, however, is the added difficulty associated with the obvious loss of control. The polymer and crosslinker slugs may not even come into contact with each other if they flood different reservoir zones/strata.

Summarizing the information presented above, the bulk gel injection gives a weak gel due to shear degradation; by contrast sequential injection gives a strong gel but has a risk of control.

Sydansk [26] proposed a gel strength classification code from A to J. Code B in his scale has a high flow behavior and is defined as a weak gel. Therefore, an alternate rheological parameter – the storage modulus (G') – could be used for gel strength evaluation. The authors classified a weak gel when G' is less than 1 dyne cm^{-2}.

Because weak polymer gels are commonly used when it is important to keep gels or gel particles as a flowing fluid, it is referred to as a flowing gel process.

Gels with high flow behavior easily propagate into high permeability zones and, in the subsequent flooding, the depth of the propagation could be adjusted. The mechanism of this process is based on fine gel particle migration through the porous media. Some of the particles pass through the throats and keep propagating forward, while others are trapped and block the flow. The resulting flow block diverts the subsequent fluid injection into low permeability zones.

The following presents an experimental study of nanogel, which allows combining the advantages of gels used in bulk and in sequential injection. The addition of light metal nanoparticles additionally increases the gel system strength via crosslinking, but has no effect on the flow ability. During the injection, very weak gel forms near the wellbore region and continues to propagate into the reservoir. Adjustable crosslinking time allows an opportunity to be sure that the gel bank is placed exactly in the thief zone and a strong in-situ gel is not formed unpredictably. The subsequent injection of fluids, either water or chemical solutions, will redirect the predominant flow paths to an unswept reservoir zone. Overall, the process improves oil sweep efficiency by water flood (or chemical flood). It therefore leads to enhanced oil recovery (EOR). The surface of the gel is rough in nature. Thereby, as shown in reference [27], it is expected that the fluid flow over the gel surface will squeeze it. From this point of view, the compressive strength of the gel also should be taken into account while estimating the gel durability.

In the following, Kamcel 1000 grade carboxyethyl cellulose (CMC) (JSC Karbokam, Russia), with a degree of substitution of carboxymethyl groups 80–90 and a degree of polymerization of 1050–1150, was used as a polymer. A polyvalent metal salt has been used as a crosslinking agent. The crosslinking time of the polymer solution has an adjustable range from 1 hour up to 10 days.

12.1.1 Nanogel Strength Evaluation

A nondestructive analysis for gel strength evaluation has been undertaken. This is rather opposite to the destructive methods when various kinds of so-named penetrators are applied, like texture analyzers and shearometers [28]. Destructive methods measure gel strength mechanical loading and is terminated when the gel fails. The principal limitation of distractive methods is that they allow only the ultimate system strength to be measured. Nondestructive methods rely on interaction between the gel and electromagnetic waves [29] or ultrasonic signal transmission [30]. The advantage of the present methods lies in the ability to monitor the dynamic of the gelation process without destruction of the test system. The methods also allow an objective valuation of the gel's compressive strength to be obtained.

The tests were conducted as follows: in a 5% aqueous solution of polymer (CMC) with permanent stirring were consequently added nanoparticles and a crosslinking agent. To avoid the effect of entrapped air, 3.4 MPa (500 psi) pressure was supplied. Tests were conducted at 62 °C. The gel with a concentration of nanoparticles of 0.0125 wt% was selected for the experiments due to the increase of gel strength up to 65% (here and after referred to as nanogel). An analysis of current research methods of polymer gels' mechanical strength testifies that by nature of the effect on a specimen they are conventionally divisible into two groups: destructive and nondestructive [31].

In view of the above, we have conducted a series of experiments on UCA by Chandler Engineering, with a view to make a sensitivity analysis of nanoparticle additives when forming polymer gel. The ultimate gel strength values obtained were verified on a shearometer produced by Fann. The results were in agreement for both methods within a 5% error interval.

A study and comparison of gelation curves in the presence and absence of nanoparticles allows the process mechanism to be explained. As seen in Figure 12.1, the process of a gel

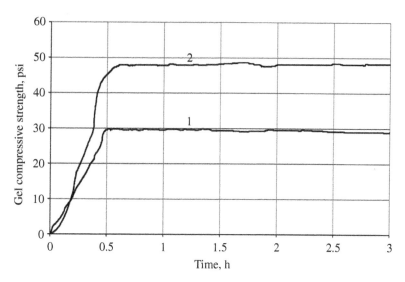

Figure 12.1 Gel compressive strength changes: 1 – gel without addition of nanoparticles; 2 – gel contains 0.0125 wt% of nanoparticles.

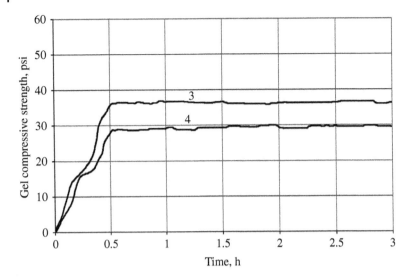

Figure 12.2 Gel compressive strength changes: 3 – gel contains 0.006 wt% of nanoparticles; 4 – gel contains 0.02 wt% of nanoparticles.

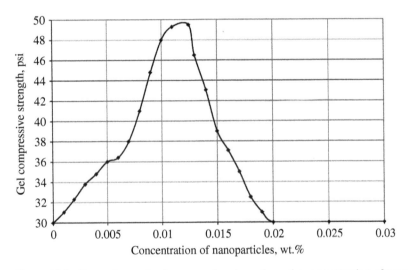

Figure 12.3 Dependence of gel compressive strength on the concentration of nanoparticles.

strength gain dynamic depends on the nanoparticle presence in the system. In the presence of nanofillers, we can visually identify an inflection point on the graph (see Figures 12.1 and 12.2). This allows us to speak about two simultaneous processes: crosslinking of the polymer system and "reinforcement" of the gel by the metallic nanoparticles [32]. We propose the following kinetic mechanism of gelation in the presence of nanofillers. Up to the concentration 0.0125 wt % nanoparticles are uniformly distributed within the gel, thus increasing the surface area of the filler, which is accompanied by a gel strength increase. However, after exceeding a threshold concentration (0.0125 wt%) the gel strength decreases (see Figure 12.3). This can be explained by aggregation of nanoparticles (and possibly coalescence) that leads to a decrease of the nanofiller surface area, which finally reduces the "reinforcing" effect [33]. To confirm our assumption we need to prove the existence of an inflection point on the gelation curve in the presence of nanoparticles in the system.

We have built an algebraic equation describing the nonlinear function shown in Figure 12.1 (line 2) in a nanofiller concentration of 0.0125 wt% in order to determine the coordinates of the inflection point and to prove its availability [34] (see Appendix B).

Relevant inflection point coordinates of the curves reflecting the gelation at different concentrations of nanofillers are displayed as 0.02 wt% $t = 0.355756$ and 0.006 wt% $t = 0.363811$. In the absence of a nanofiller in the gel, the gelation curve is straight and the inflection point is not observed.

12.1.2 Kinetic Mechanism of Gelation

Presented estimates confirm the above proposed kinetic mechanism of gelation in the presence of nanofillers. The proposed kinetic mechanism of the observed phenomena should reflect the electrical properties of the gel [35]. Aluminum nanoparticles are good conductors and their uniform distribution in the gel should increase specific electrical permeability. We suppose that the dependence of electrical permeability of the gel on nanoparticle concentration will not be monotonic. Any excess of the threshold filler concentration should decrease electrical conductivity because of uneven nanoparticle distribution.

Electrical conductivity of the gel was studied using a digital milliohm meter. Measurement accuracy in the current device accounts for 0.03%. The gel was placed in a hollow tube, with circular platinum electrodes ($R = 0.4$ cm) attached to its ends in order to measure electrical conductivity. The electrode contacts were connected to a power supply with a constant 1 V output. Freshly prepared gel was placed in a hollow tube in order to take the measurements. The distance between the electrodes was 0.8, 5.6, and 6.9 cm. Electrical conductivity was measured at intervals of 30 or 10 seconds for three hours. Figure 12.4 displays the

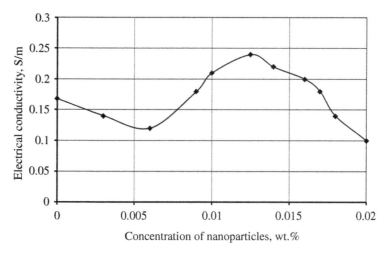

Figure 12.4 Dependence of electrical conductivity on the concentration of nanoparticles.

results of the conducted tests. Electrical conductivity of the gel is nonmonotonic, reaching its maximum at the nanoparticle concentration of 0.0125 wt%.

A percolation theory formulated for a continuous medium [36] is the one best suited to describe electrical properties of the composite material. According to the theory, each point in space with the probability $p = v_f$ meets conductivity $\sigma = \sigma_f$ and the point with probability $1 - p$ meets conductivity $\sigma = \sigma_m$ [37]. Index f is the filler and index m is for the matrix. The ercolation threshold (v_f^*) in this case is equal to the minimum percentage of space occupied by conductive areas where the system is still conducting. When changing v_f in the range of 0 to 1, electrical conductivity of the composite increases from σ_m to σ_f.

An increase in σ is nonmonotonic: most abrupt changes usually occur in a narrow range of filler concentrations (see Figure 12.4), which suggests an insulator–metal transition, or the percolation transition as it is also called, at v_f equal to the percolation threshold [38]. Let us consider the distribution of conductivity in the system at various contents of filler as v_f. At low v_f all conductive particles are combined into clusters of finite size, isolated from each other. With increasing v_f the average cluster size increases and in the case of $v_f = v_f^*$ many of the isolated clusters merge into a so-called infinite cluster, penetrating the whole system: then the conductance channel appears. A further increase in v_f leads to a sharp increase in the volume of an infinite cluster. It grows by absorbing the finite clusters, especially the biggest ones. As a result, the average size of finite clusters decreases. Researchers have noted the changes it causes in the topology of the gel, which form a so-called "dead end" [39]. Dead ends are parts of the cluster combined with the core ("electrically conductive") part of the cluster) by a single node (the connection side of the cell), which constitute a large part of the cluster but do not participate in the conductivity. This explains its decrease after reaching the threshold concentration. This concentration corresponds to the point at which the maximum gel strength is observed as a result of a uniform distribution of nanoparticles. However, a further increase in concentration leads to aggregation and coagulation. Incurred "centers" of aggregation adversely affect the gel strength because of changes in the topology homogeneity, which is reflected in the electrical conductivity due to the reasons mentioned above.

12.1.3 Core Flooding Experiments

At the existing Shallow Water Guneshli (SWG) oil field, conditions were selected to run core flooding experiments. The SWG is located at the south-eastern part of the Azerbaijani sector of the Caspian Sea. Production began in 1977. The main properties of the SWG are given in Table 12.1. Artificial core models were built by packing standard quartz sand of specific size ranges (Table 12.2). Live crude oil and formation brine were used for all the core floods. The properties of the oil and formation brine are given in Tables 12.1 and 12.3. For a secondary injection, synthetic Caspian Sea water was used. The composition of the water is given in Table 12.4.

The resistance factor, residual resistance factor, and their change trend to the test time are shown in Figure 12.5. The standard testing procedure was used during the experiments. Tests were conducted at 62 °C. Artificial core samples with 470 md permeability were used for experiments.

Table 12.1 SWG oil field properties.

Gravity (API)	Oil saturation (%PV)	Formation type	Saturation pressure (MPa)	Net thickness (m)	Average permeability (mD)	Average porosity (%)	Temperature (°C)	Initial reservoir pressure (MPa)	Current reservoir pressure (MPa)
32	78	(sand, sandstone, claystone)	23.3	66.5	195	27	62	33.4	16

Table 12.2 Artificial core parameters.

Layers	Geometric size L × W × H (cm)	Permeability (mD)	Porosity (%)	Oil saturation (%)
High	36 × 4 × 4	897	24.8	76.7
Low	36 × 4 × 4	168	21.1	72.2

Table 12.3 Crude oil and brine properties.

Properties	Crude oil	Brine
Viscosity in reservoir conditions (cP)	0.96	0.48
Atmospheric viscosity at 20 °C (cP)	4	1.001
Density (kg m^{-3})	705	1075

Table 12.4 Formation brine and sea water properties.

Water type	Density at 20 °C (kg m^{-3})	Na$^+$K$^+$	Ca	Mg^{2+}	Cr	SO$_4$ (%)	HCO$_3$	CO$_3$	RCOO	HB$_4$O$_7$
Synthetic Caspian sea water	1008.7	33.20	4.22	12.58	33.87	14.73	0.8	0.37	0.17	0.05
Formation brine	1075.1	49.01	0.22	0.77	36.13	0.95	7.95	2.95	1.6	0.43

Two different artificial core samples with quartz sand were used to reproduce a confined homogeneous or a simply stratified heterogeneous porous media (Table 12.2). The dry core samples were evacuated and saturated with formation brine under vacuum by standard procedures. Porosity and pore volume were calculated from weight differences between wet and dry cores. Each core was mounted in a core holder and at least 15 PV of the same formation brine was pumped through the core to reach a stabilized pH at the outlet and to

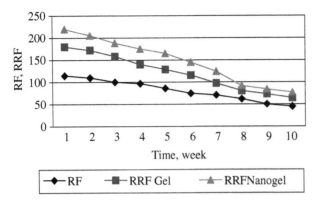

Figure 12.5 Relationship between RF, RRF, and test time.

stabilize the pressure drop across the core. After determining the permeability of the formation brine by the porous plate method [40], the residual water saturation was 14%. Then the core sample is placed back in the core holder, with simulated reservoir conditions: formation pressure 16 and 26 MPa overburden pressure and temperature 62 °C. The core sample was saturated with the live crude oil recombined from the separator oil and synthetic separator gas. After achieving a constant differential pressure, the oil permeability of the sample was measured. The core samples were aged in live crude oil for a period of two weeks. During the aging period the live crude was replaced by fresh crude oil three times. A minimum of two pore volumes was injected and a sufficient amount was used to achieve a constant GOR.

The modeling study presented in reference [41] shows that oil recovery is significantly greater if the flow is blocked in the neighborhood of the producing well as compared with blocking near an injection well or deep in the reservoir. It also shows that the crossflow between the layers of a layered reservoir plays an important role in achieving a gain in oil recovery. It is obvious that core treatment experiments without crossflow between the core samples could not accurately represent the fluid deep diversion. There in the gel bank act part more as a plunger than a blockage system because of its impermeability. Therefore, treatment experiments with crossflow between the core samples were conducted in the experimental setup shown in Figure 12.6. It has an injection vertex but separate production lines, which allows accurate monitoring of the oil produced.

Unsteady state displacements were performed at full reservoir conditions and a nominal flow rate of 5 PV per day. All experimental plans and procedures were as follows:

1) Saturating brine and experimental oil
2) Synthetic Caspian Sea water flooding up to 98% water cut
3) Gel flooding with 10% PV
4) Subsequent water flooding to 98% water cut

Six flooding experiments were with gel and nanogel at different gel bank in-situ positionings.

To test the impact of gel bank positioning on the diverted performance of the gel, three different positions of gel bank were selected (Figure 12.7). The volumes of produced water

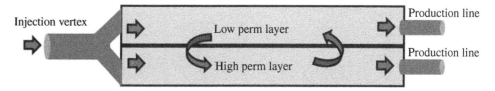

Figure 12.6 Artificial core crossflow scheme.

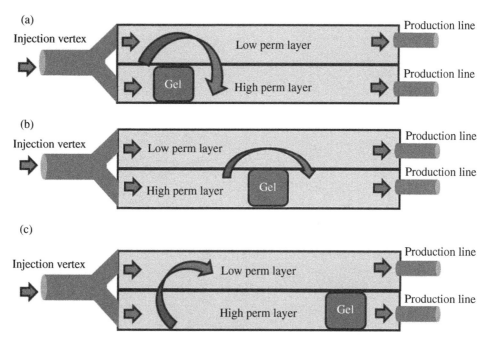

Figure 12.7 Gel bank positions. (a) P1 – position near the injection line. (b) P2 – position at the middle between the injection and production lines. (c) P3 – position near the production line.

and oil were recorded at intervals, and the oil recovery factor was calculated as a percentage of the original oil in place (%OOIP).

Experimental results show that the residual resistance factor (RRF) for nanogel application is larger in comparison with gel without nano additives and has a more effective life-period. Nanoparticle addition increases RRF of polymer gel by up to 22% at the first week and keeps this without a significant drop even after 10 weeks. The resistance factor (RF) increase was 56 and 91.3% respectively. Figure 12.5 presents the RF and RRF test results.

These results were also confirmed during core flooding. As shown in Table 12.5, the nanogel treatment leads to 10.7% increase of oil recovery from a low permeable layer and allows a 69% OOIP of total recovery. Gel treatment results showed a 63% OOIP of oil recovery and only a 4.5% recovery increase from the low permeability layer (Table 12.6). Thereby it is obvious that in deep fluid a diversion takes place for both cases.

Table 12.5 Nanogel flooding experiment results.

End of flooding	Water cut (%)	Oil recovery (%)	Volume percent (%)	
			H	L
Caspian water	96.8	35.2	79.2	20.8
Nanogel	87.7	37.6	76.1	23.9
Caspian water	96.7	69	68.5	31.5

Table 12.6 Gel flooding experiment results.

End of flooding	Water cut (%)	Oil recovery (%)	Volume percent (%)	
			H	L
Caspian water	99.4	34.2	81.2	18.8
Gel	86.7	38.2	76.1	23.9
Caspian water	96.7	63	74.4	27.6

Table 12.7 Gel flooding experiments with different gel bank positions.

Gel bank position	Gel			Nanogel		
	Oil recovery (%)	Volume percent (%)		Oil recovery (%)	Volume percent (%)	
		H	L		H	L
P1	58	83.9	16.1	59	82.1	17.9
P2	60	78.1	21.9	64.3	76	24
P3	63	74.4	27.6	69	68.5	31.5

In order to test gel position sensitivity on oil recovery three different positions of the gel bank were selected, as shown in Figure 12.7. Table 12.7 demonstrates that the position P1 (i.e. near the injection line) weakly affects oil recovery. This does not depend on the gel type we used and stays the same for both of them. This is expected as the crossflow between layers leads to fluid diversion into the high permeability layer after the blockage. The position P2 (i.e. at the middle between the injection and production wells) is more preferable than P1 and leads to 12% of OOIP and 13.3% of OOIP recovery increase correspondingly. However, the position P3 is the most efficient and leads to 19 and 28% of OOIP recovery increase in comparison with the position P1.

12.1.4 Concluding Remarks

- Nanoparticle addition increases the RRF of polymer gel up to 22% in the first week and above 20% after 10 weeks as compared with polymer gel. The resistance factor increases were 56 and 91.3% respectively.
- The oil recovery increase for a nanogel application was 6% OOIP in comparison with gel without nanoparticles. The total oil recovery was 69% OOIP.
- In a depth fluid diversion takes place for all gel compositions, but for nanogel the result was more profound and long lasting. The nanogel treatment leads to a 10.7% increase in oil recovery from a low permeability layer.
- The gel bank position is very critical for gel treatment. The flow blockage leads to a 19 and 28% OOIP recovery increase if the gel bank is placed near the production well for both gel and nanogels respectively.

12.2 Colloidal Dispersion Nanogels

Application of polymer solutions is a separate branch of EOR methods [42–45]. The two major subdivisions of polymer influence mechanisms are sweep efficiency improvement and IFD [46–48]. Consequently, polymers are utilized as aqueous solutions and gels [49–51]. Addition of nanomaterials to the mentioned polymer systems significantly change polymer system physical–chemical properties [52–54]. The following is related to the nano-treated CDGs for in-depth permeability modification. CDGs in comparison with bulk gels have missing continuous intermolecular gel networks, which are predominantly due to the low concentration of polymer in their composition [55, 56]. The system in essence is a microscale uniformly separated dispersed gel/colloid [57, 58], formed by a crosslinking reaction that reduces the motional degrees of freedom compared to a pure polymer [59].

From the first pilot project until the present the CDG application underwent a long evolution in chemical compositions, treatment mechanisms, and gained notable experience [60–62]. However, to date it still suffers from some serious drawbacks [63, 64]. The propagation volume and strength of the formed immobile gel structure are still highly debated topics among researchers and practitioners [65–67].

The absence of standardized study and research methods on CDG confuse the comparison of different results and findings. It is very difficult to make a sensible comparison when taking into account the differences in temperature, reaction mechanisms, polymer chemistry, polymer/crosslinker ratio, to name a few of the variables [68, 69].

The following presents the results of laboratory studies on the CDG nanogel. Experiments were done in measuring sand pack flooding, IFT, rheology, and the zeta potential.

2-Acrylamide-2-methylpro-pane sulfonic acid (AMPS), acrylic acid (AAc), hydrolyzed polyacrylamide (HPAM), and chromium triacetate as a metallic cross-linker, which were sourced from Sigma-Aldrich. The initiator (4,4′-azobis-(4-cyanovaleric acid), ACVA), the inhibitor (hydroquinone), and sodium hydroxide were also obtained from the same supplier. Nitrogen gas used for degassing solutions was supplied locally and had 99.999% purity. The solid anatase/rutile TiO_2 nano powders were purchased from the US Research Nanomaterials Inc. (USA) and information on the nano powders is provided in Table 12.8.

Table 12.8 Properties of TiO$_2$ nanoparticles.

Purity	APS (D50)	SSA	Bulk density	PH	Loss of weight in drying	Loss of weight on ignition
99+ %	20 nm	10–45 m^2 g^{-1}	0.46 g ml^{-1}	5.5–6.0	0.48%	0.99%

Firstly, 0.05 wt% of nano-sized titanium dioxide (titania, TiO$_2$) was dispersed in water with an anionic surfactant in an ultrasonic bath. The bath was water cooled at a constant temperature, 25 °C, in order to prevent overheating of the dispersion. Then monomer solution of AMPS and AAc were prepared, so that the total monomer concentration of solution was equal to 1 M. Afterwards, sodium hydroxide was added into the solution to adjust the pH to approximately 7 (±0.5). Then 0.005 M initiator (ACVA) and sodium chloride were added to ensure constant ionic strength among the experiments. At the next step the solutions were purged with nitrogen for two hours. After degassing, in a temperature-controlled bath at 40 °C free-radical polymerizations of the solution (aqueous phase) were carried out. After approximately 24 hours 1 ml of 0.2 M hydroquinone solution was added to stop the polymerization. Then the solution was mixed with an HPAM solution and a dispersed TiO$_2$ nanoparticle suspension was added to this solution. At the last preparation step, the aqueous solution of chromium triacetate was added. The solutions were stirred at the gelation temperature using a magnetic stirrer for 24 hours and left for up to eight days at the ambient temperature to ensure an adequate crosslinking.

12.2.1 Rheology

A minor decrease in viscosity has been observed for a week. After this, the viscosity of all tested solutions stabilized and remained constant. This is explained by intramolecular crosslinking of polymer chains occurring at low polymer concentrations. Small quantities of polymer in CDG compositions avoid creating a three-dimensional network with intermolecular crosslinking bonds. Evidently, the crosslinking reaction continued for up to seven days followed by polymer coil formation. The observed trends were similar for all solutions [49, 66] (see Figure 12.8).

Viscosity dependence on the crosslinker concentration was evaluated. Higher crosslinker concentration is in an inverse relation with CDG viscosity due to colloid collapse (e.g. an increase in coil number) (Figure 12.9). Presence of nanoparticles in CDG composition leads to less viscosity reduction. Most probably, titanium dioxide nanoparticles functioned as an inorganic initiator and crosslinker. The pendant groups of the polymer chains form strong networks for the polymerization.

The major challenge of polymer application in EOR is harmful impact of mineralization on the polymer structure. Mineralization quite often leads to insoluble precipitation and dramatically reduces the permeability. In this regard, the rheological behavior of 1000 ppm nanoparticle compositions has been studied at different salinities (Table 12.9).

The collapse of polymer coils in a high-salinity environment resulted in a viscosity decrease for all samples. As expected, nano-CDG due to a strong polymer network initiated

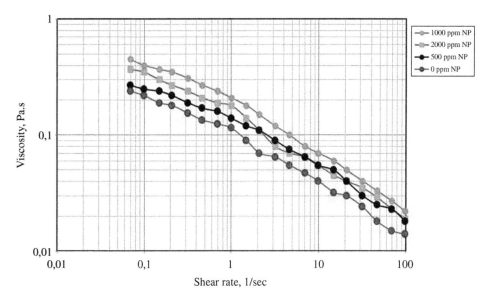

Figure 12.8 Viscosity–shear rate behavior for CDG nanogel versus nanoparticle concentration, 62 °C.

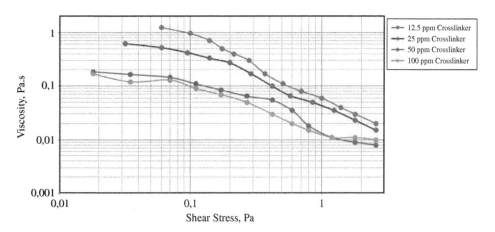

Figure 12.9 Viscosity–shear behavior for CDG nanogel versus crosslinker concentration.

by the presence of titanium dioxide showed higher durability. However, the impact of nanoparticle concentration is not uniform. The concentration of nanoparticles is in a direct relationship with mineralization durability of nano-CDG up to concentrations at around 1000 ppm. A further increase in nanoparticle content showed the formation of a weak bulk gel. This was explained by the crosslinking behavior of the nanoparticle titania. Small concentrations of polymer in solution do not allow the formation of a strong polymer matrix. This phenomenon is out of the scope of the current study and will be the topic of future investigations.

Table 12.9 Effect of salinity on the viscosity of 1000 ppm nanoparticle CDG systems as a function of crosslinker concentration (shear rate: $1\,s^{-1}$, after seven days of preparation at 25 °C).

Water salinity (ppm)	Crosslinker concentration (ppm)	Viscosity (Pa.s)
0	0	0.251
0	25	0.123
0	50	0.0341
2000	0	0.203
2000	25	0.0172
2000	50	0.0054
4000	0	0.0113
4000	25	0.0067
4000	50	0.0038
8000	0	0.0022
8000	25	0.0024
8000	50	0.0017

12.2.2 Aging Effect

It is obvious that the reservoir condition creates a destructive environment for polymer systems. The dispersed phase of CDG and the absence of a continuous polymer network prevents polymer coils from degradation. The impact mechanism of the first factor is dilution of detrimental electrolyte media of a connate water. The second effect is based on the small contact area of polymer coils in comparison with a bulk gel [70].

The aging experiment shows low CDG degradation in the presence of nanoparticles, regardless of salinity. Addition of nanoparticles probably increases the stability of the obtained gel due to the formation of the nucleus inside polymer coils. Meanwhile, in the case of nanoparticles, the absence of a processing crosslinking reaction is associated with a decrease in the polymer coil size. From that point of view, nanoparticles play a role of "a brake" in crosslinking reactions and limit further depletion of the polymer coil size.

The association of negative charges along the polymer chain with positive charges from the surrounding media frequently leads to precipitation. At the end of the experiment the precipitation was not observed in any of the studied samples [57, 63] (Figures 12.10 and 12.11).

12.2.3 Interfacial Tension

The addition of nanoparticles reduces cohesive energy at the liquid–air interface, which causes a decrease in the base fluid surface tension (Table 12.10). The observed phenomena are related to the nanoparticle's Brownian motion in the liquid. Consequently, the lower level of interface total free energy has been achieved due to the nanoparticle dispersal at

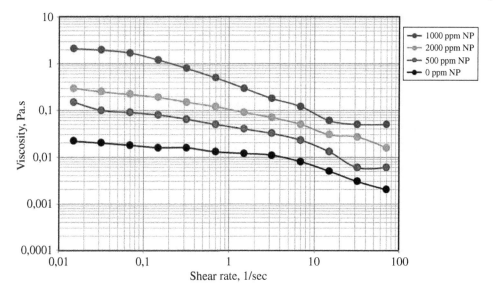

Figure 12.10 Effect of thermal aging in distilled water on shear viscosity (aging temperature 90 °C, aging time 30 days).

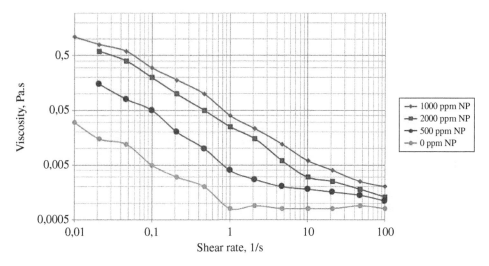

Figure 12.11 Effect of thermal aging in electrolyte media on shear viscosity (aging temperature 90 °C, aging time 30 days).

the liquid–air interface [71, 72]. At low concentrations, the polymers also show the surfactant behavior attributed to the same effect (e.g. a reduction of the interfacial energy). Hence, both of the described phenomena lead to the scaling down of surface tension values. However, exceeding nanoparticle critical concentrations (at around 1000 ppm) has the opposite effect, which is associated with the effect of Waals–London molecular forces and the rising electrostatic force of repulsion. This means that the intermolecular attractive forces of

Table 12.10 Interfacial tension values of CDG versus nanoparticle concentration.

Nanoparticle concentration (ppm)	IFT (mN m^{-1})
0	40.74
500	29.67
1000	19.34
2000	26.12

Table 12.11 Zeta potential of CDG/nano-CDG solutions.

	Zeta potential (mV)				
Description	1st measurement	2nd measurement	3rd measurement	Mean value	Standard deviation
CDG without nanoparticle	−31.6	−32.4	−34.1	−32.7	1.276
CDG with 1000 ppm nanoparticle	−48.7	−46.4	−46.1	−47.06	1.422
CDG with 2000 ppm nanoparticle	−27.1	−24.8	−23.5	−25.13	1.823

nanoparticles dominate repulsion, which results in nanoparticle agglomeration. To prove this statement the following series of experiments have been performed.

12.2.4 Zeta Potential

Zeta potential values above ±30 mV are regarded as a good indicator of colloid stability. From the two functional groups of HPAM molecules (i.e. amide and carboxylic), the percentage of carboxylic (—COO—) groups is the major factor affecting anionicity. In other words, an increment in the quantity of carboxylic groups in the polymer structure results in an increase in anionicity [73, 74]. It can be seen in Table 12.11 that all CDG compositions present good stability. The addition of nanoparticles increases the anionicity of the colloid due to the negative charge of the TiO_2 nanoparticle. The nanoparticle aggregates on a polymer coil/solution interface at lower concentrations (below 1000 ppm) in accordance with IFT outcomes. Meanwhile, exceeding the concentration over 1000 ppm reduces the zeta potential value because of nanoparticle stacking without any contact with the coil surface. The anionicity is reduced due to diffusive layer shrinkage and low electrical potential of polymer coils. It is possible to assume that the nanoparticles are crosslinked in the polymer structure and increase the particle sizes in solution that cause further agglomeration. This leads to the conclusion that the optimum nanoparticle concentration is the most important factor in nano-CDG composition selection (Figure 12.12).

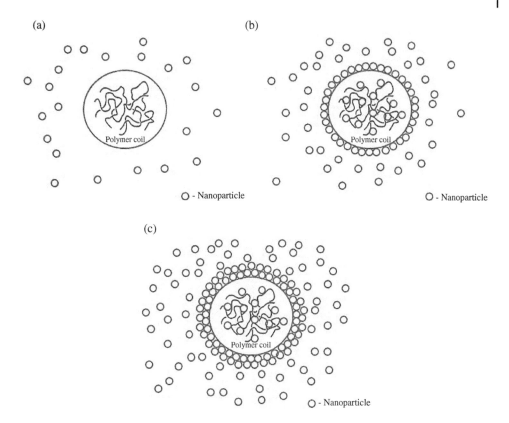

Figure 12.12 Illustration of nanoparticle aggregation on polymer coils. (A) – addition of nanoparticles. (B) –aggregation of a nanoparticle on polymer coil. (C) – nanoparticle stacking on polymer coil decreasing zeta potential.

Table 12.12 Average particle size values for CDG compositions.

	Average particle size (nm)			
Description	After 1 day	After 3 days	After 7 days	After 10 days
CDG without nanoparticle	1258	1130	946	940
CDG with 1000 ppm nanoparticle	1452	1320	1310	1308
CDG with 2000 ppm nanoparticle	1846	1782	1645	1633

12.2.5 Particle Size Distribution

The CDG particle reduces in size up to 25% during the observation time (Table 12.12). The main cause of this outcome is continuous intra-molecular crosslinking. The polymer coil size decreases during this process, due to the increment of crosslinking points. The same trend was observed in up to 1000 ppm nanoparticle concentration. In the intervening time,

Figure 12.13 Cartoon illustrating the adsorption of TiO_2 on polymer chains for a given polymer concentration: (A) particle increasingly adsorbed on to polymer chains via hydrogen bonds; (B) saturation reached, nanoparticle accumulating and bridging in the network.

the reduction was not more than 10%. The nanoparticle concentration increases by an average particle size in nano-CDG. Most probably, the interfacial deposition of the nanoparticles limits the polymer coil shrinkage. In addition, it is possible to envisage that the formation of hydrolysable covalent crosslinks between the TiO_2 nanoparticles and HPAM induced strengthening of the connection among the coils. It is assumed here that below the critical concentration nanoparticles adsorbed on to the hydrophobic chain because of the formation of hydrogen bonds between the carbonyl groups of HPAM and TiO_2. Therefore, no free titania nanoparticles are present in the bulk solution below 1000 ppm However, if the nanoparticle concentration is above this value, free TiO_2 nanoparticles become available, which can act as bridges between different polymer coils (Figure 12.13). In addition, free particles may also agglomerate and attach together via Ti−O−Ti bindings. Overall, the PSD results prove the statement mentioned above.

12.2.6 Resistance Factor/Residual Resistance Factor

All experiments were conducted in a 10 m sand-packed system with six consecutive sections (each 152.4 cm) (Figure 12.14). Pressure transducers were placed between each section and at the system outlet/inlet. Washed and dried silica grains have a size distribution of between 45 and 60 nm. Screen and glass fiber at the inlet and outlet of the sand-pack were used to avoid sand removal. Porosity had been determined using a weight method based on the difference between the mass of a dry sand-pack saturated with distilled water. The section permeability variation was not more than 10%. After the resistance factor data were collected, the system was shut down for 48 hours to allow the gel to begin to form. Next, a minimum of 3 PV seawater was injected with a low to high flow rate. The measurements were taken when the pressure drop stabilized at each flow rate and residual resistance factors were calculated.

The main purpose of CDG injection is to develop an impermeable barrier in porous media in order to divert fluid flow from highly swept areas. The next major factor is the propagation ability of popped particles because the gel bank placed near the wellbore zone is not efficient [43]. The pressure drops to move popped particles from section to section and is measured at predetermined times. The measurement period was correlated with viscosity changes of the injected fluids. The samples of each injected fluid were kept in a thermal bath and viscosity values were determined every 24 hours. Every 50% increment in viscosity was considered as the next propagation test start.

An increment of 410% was observed in a pressure drop throughout the 1000 ppm nano-CDG test. Other concentrations of nanoparticles gave a 335 and 150% increase in RF relatively (Figure 12.15).

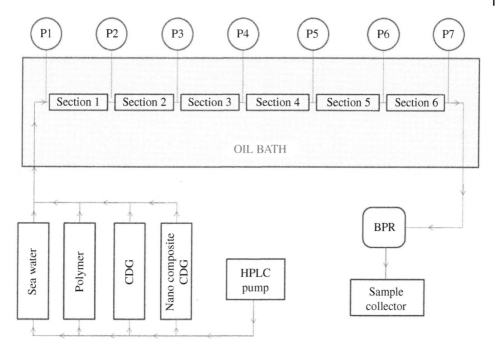

Figure 12.14 Scheme of the sand-packed flooding system.

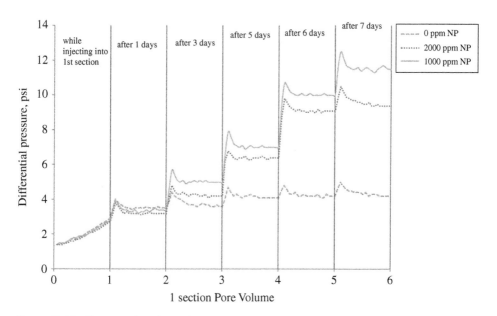

Figure 12.15 Pressure drop dependency on nanoparticle concentration in CDG.

Table 12.13 Residual resistance factor versus nanoparticles concentration.

Nanoparticle concentration (ppm)	RRF
0	524
1000	835
2000	629

Despite the higher blocking ability of 2000 ppm nanoparticle CDG (e.g. a larger particle size), the observed RF/RRF values were lower than expected. A probable explanation is that the higher critical concentration levels of nanoparticles causes a weak polymer barrier. The main reason for this is that the colloids with a low zeta potential value become unstable and collapse under the high pressure.

RRF values had been measured after washing out the CDG solution. The results are similar to earlier observations and are displayed in Table 12.13.

12.2.7 Concluding Remarks

1) Addition of TiO_2 nanoparticles to CDGs resulted in increased rheological stability, reduced interfacial tension, increased zeta potential, and a significant increment in RF/RRF values.
2) The critical concentration of nanoparticles in CDG was determined as 1000 ppm. A pseudoplastic behavior was observed with low thermochemical destruction and 35% decrease of IFT in the liquid/air interface. Meanwhile, the average particle size scale-down did not exceed 10%. The presence of nanoparticles increases the zeta potential of colloids by 43% and the RF/RRF values were 173 and 59% more than the base CDG solution respectively.
3) A kinetic mechanism of colloidal dispersed gels gelation in the presence of TiO_2 nanoparticles is proposed. It is thought that interfacial deposition of nanoparticles reduces the cohesive energy at the liquid–air interface, which causes a decrease in surface tension. Meanwhile, inter-coil deposition of nanoparticles limits polymer coil shrinkage, which was observed in particle size measurements over time. Above the nanoparticle critical concentration (e.g. 1000 ppm), the anionicity of the colloid was reduced as a result of diffusive layer shrinkage and the low electrical potential of polymer coils. Moreover, a free TiO_2 nanoparticle becomes available and acts as a bridge between different polymer coils. This results in dominance of attractive forces over repulsive forces and low zeta potential values make the colloids unstable.

13

Nanogas Emulsions as a Displacement Agent

13.1 Oil Displacement by a Newtonian Gasified Fluid

It is a well-known technique in oil reservoir development when water and hydrocarbon gas (natural gas or associated petroleum gas) are injected alternatively or simultaneously into the formation. Furthermore, the amount of injected hydrocarbon gas should not exceed its needed volume by less than 10 times for complete saturation in water at the formation pressure. Thus, the gas–water ratio reduced to the formation pressure should not be less than 2 nm^3 m^{-3} MPa while the gas–water ratio needed for water complete saturation is approximately equal to 0.2 nm^3 m^{-3} MPa under formation conditions. Hence, a significant portion of the hydrocarbon gas does not dissolve in water and is in the free gas phase state, even under high injection pressures. The shortcoming of this method is the two-phase nature of the water–gas mixture, which determines the outrunning gas inflows into the production wells. This decreases the formation flooding efficiency and oil recovery. Moreover, the application of this method requires high specific expensive hydrocarbon gas consumption. During implementation of the method, the injection well equipment becomes much more sophisticated and its injection capacity decreases no less than four times.

At the same time, the technique for oil formation flooding by water–air solutions in the pre-transition region is known. It is implemented by maintaining the ratio of air and water volumes within the range of 0.27 : 1–0.36 : 1 and the ratio of the bottom hole pressure in the injection well and formation pressure within the range of 1.1–1.8 [75]. Carbon dioxide is also used as a dispersion phase for nanogas emulsion formation. Results of a core flooding experiment were reported for North Sea high-permeable cores and show 14 and 18% oil recovery increases when the displacement was done with CO_2 and methane-based gas emulsions respectively [76].

The water–air mixture, when in the pre-transition region, increases the formation flooding efficiency. The shortcoming of the technique is intense corrosion of the oil field equipment. The air–hydrocarbon gas mixtures also increase the possibility of an explosion in the production well.

This chapter presents the experimental studies of the water–gas solution (WGS) application as a displacement agent in the pre-transition region [77] with and without surfactant additives.

Nanocolloids for Petroleum Engineering: Fundamentals and Practices, First Edition.
Baghir A. Suleimanov, Elchin F. Veliyev, and Vladimir Vishnyakov.
© 2022 John Wiley & Sons Ltd. Published 2022 by John Wiley & Sons Ltd.

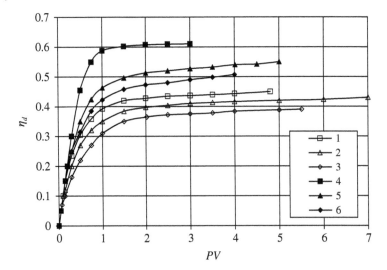

Figure 13.1 Oil recovery curves: 1, water; 2, water and WGS at $P/P_c = 2$ with sulphanol additive ($C = 0.01\%$); 3, water and WGS at $P/P_c = 2$ with sulphanol additive ($C = 0.035\%$); 4, WGS at $P/P_c = 1.3$; 5, WGS at $P/P_c = 1.3$ with sulphanol additive ($C = 0.01\%$); 6, WGS at $P/P_c = 1.3$ with sulphanol additive ($C = 0.035\%$).

The experiments were conducted on a two-layer formation model with a layer permeability of 0.4 and 1.6 µm². The porous medium was made of quartz sand (90%) and montmorillonite clay (10%). Synthetic oil was used as the reservoir oil model. The WGS with methane as the gas phase was first prepared in a PVT bomb with the saturation pressure $P_c = 3$ MPa and water–gas ratio of 1 m³ m⁻³. The experiment was conducted at 303 K. Oil displacement was carried out at a 0.1 MPa differential and various anionic surfactant concentrations (sulphanole) $C = 0$–0.4%.

The experiment results were expressed as displacement coefficient dependence on the relative pore volume of the injected agent (Figure 13.1). It is obvious that the application of the WGSs with the anionic surfactant additives significantly (13–22%) increased the displacement coefficient at an average pressure $P = 1.3P_c$. This was observed for the whole range of the cationic surfactant concentrations between 0.01 and 0.035%. The maximum displacement coefficient value is reached with displacement by the WGS without anionic surfactant additives. It exceeds the water displacement coefficient by almost 36%.

Thus, hydrophilization of the pore channels, due to anionic surfactant addition in WGS, decreases the displacement process efficiency compared to the WGS without any additives.

13.2 Oil Displacement by a Non-Newtonian Gasified Fluid

Practical applications of the gasified polymer solutions in the pre-transition region with the view to improve oil recovery efficiency during oil formation flooding are of great interest. This section shows the experimental studies of the displacement properties of the gasified polymer solutions.

The experiments were conducted on a two-layer formation model (with interfacial layers) with air permeability of 3 μm². Synthetic oil was used as the reservoir oil model. The WGS with methane as the gas phase was initially prepared in the PVT bomb with the saturation pressure $P_c = 3$ MPa and water–gas ratio of 1 m³ m⁻³. The temperature was maintained at 303 K during the experiments. Oil displacement was carried out at different pressure values in the formation model $P = 1.1–3P_c$ and various polyacrylamide (PAM) concentrations up to 0.03%.

In the first series of experiments, displacement was carried out at the polymer solution with pseudoplastic flow behavior (at the pressure differential of 0.2–0.6 MPa depending on the PAM concentration).

The experimental results were expressed as the displacement coefficient dependence on the relative pore volume of the injected agent. The obtained results are shown in Figure 13.2, where it can be seen that application of the WGSs with PAM additives allows a significant (20–36%) increase in the number of displacement coefficients at the average pressure $P = 1.3P_c$ and range of PAM concentrations of 0.01–0.03% as compared to water. The displacement coefficient has its maximum value when conducting displacement by the WGS without PAM additives and with the PAM concentration of 0.03%. In this case, it also exceeds the displacement coefficient for water by almost 36%. Furthermore, the PAM addition to the WGS only in concentrations of 0.03% is as effective as the WGS without any PAM additives.

In the second series of experiments, displacement was carried out on the polymer solution with dilatant flow behavior. The differential pressure was in a range of 0.04–0.1 MPa depending on the PAM concentration. It can be seen (Figure 13.3) that the displacement efficiency is significantly higher than in the first series. At the same time, when the PAM concentration increases, the influence of the polymer solution gas saturation on the displacement efficiency decreases since, at the concentration of 0.01%, the displacement

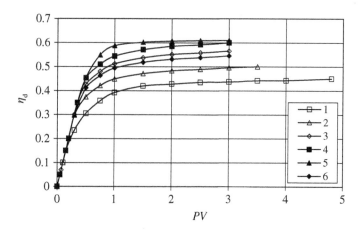

Figure 13.2 Oil recovery curves: 1, water; 2, water with PAM additive ($C = 0.01\%$); 3, WGS at $P/P_c = 1.3$ with PAM additive ($C = 0.01\%$); 4, water with PAM additive ($C = 0.03\%$), WGS at $P/P_c = 1.3$ with PAM additive ($C = 0.02\%$); 5, WGS at $P/P_c = 1.3$, WGS at $P/P_c = 1.3$ with PAM additive ($C = 0.03\%$); 6, water with PAM additive ($C = 0.02\%$).

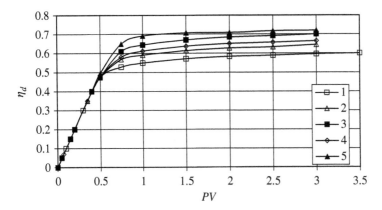

Figure 13.3 Oil recovery curves: 1, water with PAM additive ($C = 0.01\%$); 2, water with PAM additive ($C = 0.02\%$); 3, water with PAM additive ($C = 0.03\%$), WGS at $P/P_c = 1.3$ with PAM additive ($C = 0.02\%$); 4, WGS at $P/P_c = 1.3$ with PAM additive ($C = 0.01\%$); 5, WGS at $P/P_c = 1.3$ with PAM additive ($C = 0.03\%$).

coefficient for the gasified solution is 11% higher; at $C = 0.02\%$, it is 8% higher and at $C = 0.03\%$, it is 3% higher than for the degassed one.

The positive effect is explained by the known phenomena seen during the gasified polymer solution displacement process, where the two following processes occur:

- The PAM concentration increment leads to injection fluid viscosity growth.
- The slippage effect is reduced as the porous medium wettability increases.

Meanwhile, application of the gasified polymer solution reduces polymer consumption. For instance, displacement efficiency of WGS with 0.02% PAM is as effective as a 0.03% PAM solution injection.

13.3 Mechanism of Observed Phenomena

As shown above, the slippage effect strongly correlates with the average radius of the pore channel. Since the flow profile of the gas-fluid system is near (above) the saturation pressure in the heterogeneous porous medium, it should be more uniform compared to the degassed Newtonian fluid at $P \gg P_c$ [78, 79].

Indeed, consider the flow of a gasified fluid near the saturation pressure that takes place in two parallel layers with the permeability of k_1 and k_2, when $k_1 \gg k_2$. The average radius of the pore channels in each layer can be determined from the known ratio $R_{mi} = 2\sqrt{2k_i/m_i}$ ($i = 1,2$). The flow rate and apparent viscosity during the flow with slippage are determined for each layer from the expressions

$$Q_i = \frac{k_i F}{\eta_{Si}} \frac{\Delta P}{l}; \quad \eta_{Si} = \frac{\eta_0}{\left(1 + 2b\sqrt{\frac{m_i}{2k_i}}\right)}$$

The mobility ratio at the same pressure gradients is

$$\frac{k_1}{\eta_{S1}} : \frac{k_2}{\eta_{S2}} = \frac{Q_1}{Q_2} = \frac{k_1\left(1 + 2b\sqrt{\frac{m_1}{2k_1}}\right)}{k_2\left(1 + 2b\sqrt{\frac{m_2}{2k_2}}\right)}$$

Since $k_1 \gg k_2$, the value within the brackets in the denominator of the specified formula is always bigger than the one in the numerator. For example, if $k_1 = 10^{-12}$ m²; $k_2 = 4.1 \times 0^{-14}$ m²; $m_1 = 0.25$; $m_2 = 0.2$; $b = 10^{-6}$ m, the mobility ratio will be equal to 10.3 instead of 25 without the slippage. Thus, the slippage effect leads to filtration profile alignment of the gasified fluids near (above) the saturation pressure.

These conclusions are confirmed by the experimental data. It should be noted if the low-permeable layer thickness increases then the impact of this phenomena will be more significant. Indeed, the sweep efficiency in the low-permeable interlayer with the water influx from the high-permeable formation (i.e. the part of the formation advanced by the front in the low-permeability layer) is equal to

$$\eta = \frac{1 - \sqrt{1 - 2(1 - \eta_{22})\beta}}{1 - \eta_{22}}$$

where

$$\beta = \frac{k_2}{k_1}\frac{\eta_1}{\eta_2}\frac{m_1}{m_2}\frac{1+\eta_{11}}{2}, \quad \eta_{11} = \frac{\eta_w}{\eta_1}, \quad \eta_{22} = \frac{\eta_w}{\eta_2}, \quad \eta_i = \frac{\eta_0}{\left(1 + 2b\sqrt{\frac{m_i}{2k_i}}\right)}$$

Here $\eta_w, \eta_1, \eta_2, \eta_0$ are water viscosity, apparent oil viscosity with the account of the slippage for the high- and low-permeability layers, and the real oil viscosity respectively. For the case with slippage the following values are valid: $k_1 = 10^{-12}$ m², $k_2 = 10^{-13}$ m², $m_1 = 0.205$, $m_2 = 0.17$, $\eta_w = 10^{-3}$ Pa s, $\eta_0 = 2 \times 10^{-3}$ Pa s, and $b = 10^{-6}$ m. In this case, for the sweeping efficiency, one obtains $\eta = 0.184$ ($\eta_1 = 1.22 \times 10^{-3}$ Pa s, $\eta_2 = 7 \times 10^{-4}$ Pa s). For the case without slippage, when $\eta_1 = \eta_2 = 2 \times 10^{-3}$ Pa s, $\eta_{11} = \eta_{22} = 0.5$, the sweeping efficiency drops to $\eta = 0.092$. It is possible to see that the sweeping efficiency for the case with slippage is two times higher than without it.

Considering the influence of the high- and low-permeable layer thickness on the filtration profile, the ratio of the hydraulic permeabilities could be expressed as follows:

$$\frac{Q_1}{Q_2} = \frac{k_1 h_1}{\eta_1} \bigg/ \frac{k_2 h_2}{\eta_2}$$

Taking the parameters of the previous analysis, as well as $h_1 > h_2$, when $h_1 = 20$ m and $h_2 = 10$ m, and for the case $h_1 < h_2$, $h_1 = 10$ m, $h_2 = 20$ m, and accounting for slippage, one can get

$$\text{for } h_1 > h_2, \quad \frac{Q_1}{Q_2} = 11.48$$

$$\text{and for } h_1 < h_2, \quad \frac{Q_1}{Q_2} = 2.87$$

It can be seen from the obtained results that when the layer with lower permeability, where the slippage effect is more pronounced, has a large thickness, the effect of the flow profile alignment is more significant. Furthermore, at a certain value of h_2/h_1 (in the considered case, at $h_2/h_1 > 5.74$), the flow profile becomes less uniform since the flow rate of the low-permeable interlayer starts to exceed the flow rate of the high-permeable one. In this case, it is necessary to regulate the slippage coefficient with the help of either surfactant additives or an injection pressure change.

Similarly, without slippage, one can see that in the case

$$h_1 > h_2, \quad \frac{Q_1}{Q_2} = 100$$

and in the case

$$h_1 < h_2, \quad \frac{Q_1}{Q_2} = 5$$

In other words, when the low-permeable layer thickness increases, the effect of the flow profile alignment enhances since, when the thickness increases, the flow rate grows. However, the effect is much lower than the one for the case with slippage (almost 10 times lower at $h_1 > h_2$ and two times lower at $h_1 < h_2$). At the certain value of h_2/h_1 (in the considered case, at $h_2/h_1 > 10$), the flow profile becomes less uniform since the flow rate of the low-permeable layer starts to exceed the flow rate of the high-permeable one.

It should be noted that application of the gasified and initially dilatant polymer solutions in the pre-transition phase state acting as the displacement agent are also promising. Furthermore, all the issues associated with polymer flooding are determined. Indeed, formation of gas bubbles on the porous medium surface (in accordance with and at the oil–gas ratio of 0.1–10, the porous medium surface is covered by the gas phase bubbles) prevent the polymer solution from adsorption. The dilatancy and slippage effect contributes to the flow profile alignment, as well as to the acceleration of the overall process development and increase in the well injection capacity due to the flow modification from the dilatant to the pseudoplastic. There is also a possibility to use the polymer solutions, which have a pseudoplastic flow behavior, because the slippage effect neutralizes the shear thinning impact on the flow profile.

Let us consider the filtration of the gasified polymer solution in the pre-transition region that takes place in two parallel layers with the permeability of k_1 and k_2 (all other conditions are equal). Moreover, $k_1 \gg k_2$ and the process obeys the exponential flow law with slippage (see Section 4.2.2. Then, from Equation (2.72),

$$\frac{Q_1}{Q_2} = \frac{k_1}{k_2} \left(\frac{k_1}{k_2}\right)^{\frac{1-n}{2n}} \left(1 + \frac{3n+1}{n} \frac{b\sqrt{m}}{\sqrt{8k_1}}\right) \left(1 + \frac{3n+1}{n} \frac{b\sqrt{m}}{\sqrt{8k_2}}\right)^{-1}$$

It is obvious that the slippage enhances the effect of the flow profile alignment for the dilatant fluid ($n > 1$) and neutralizes the shear thinning impact on the flow profile of the pseudoplastic fluid ($n < 1$).

To sum up, it can be concluded that, during application of gasified fluids in the pre-transition region, accounting for the slippage effect explains the front alignment and growth of the displacement coefficient.

13.4 Field Application

The nanogas emulsion injection has been commercially applied in the Bibiheybat oil field in Azerbaijan from 1988 to 1990. The oil field has been developed since 1890, with water flooding since 1954 and a disordered injection/producer design.

A typical facility arrangement used for nanogas emulsion injection is shown in Figure 13.4. The technology does not require any special technical capacities and could be implemented as part of a basic waterflooding facility. In the particular case, an air supply has been provided by the existing compressor facilities for the airlift operations. For the field trial 10 injection wells were used (in particular, seven injectors for horizon V and three injectors for horizon VII). This arrangement provided covering for 65–68 production wells (53–55 in horizon V and 12–13 in horizon VII).

The following result has been achieved:

- Horizon V – the injection water volume had decreased in a range of 15–20%. In the meantime, oil and water flow rates increased by 10–12%.
- Horizon VII – produced water decreased by 8–10%, but oil recovery values did not change.

In the field the nanogas emulsion injection resulted in an increase of an entire field oil recovery in a range of 8–10% and in 10–15% decrement in injected water volume. Figure 13.5 presents the oil production dynamics during the field trial.

It was observed that the water injection rate had to increase sharply by 20–40% with simultaneous air injection (note that the air/water ratio was 0.6–1.2 m^3 m^{-3}) and distinctly decreased up to the base level after the termination of water injection or while exceeding the

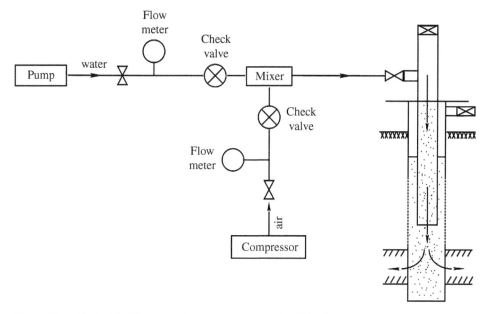

Figure 13.4 Typical facility design for a nanogas emulsion injection.

214 *13 Nanogas Emulsions as a Displacement Agent*

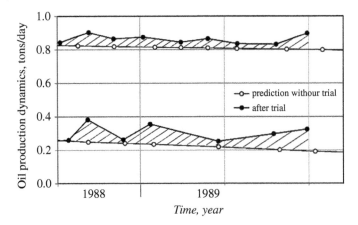

Figure 13.5 Oil production dynamics.

air/water ratio to more than 1.5 m³ m⁻³. The water injection dynamics are presented in Figure 13.6.

Nanogas emulsion injection has also been applied in the Lyantor oil field, in West Siberia (Russia), unit AC-9. The base system of reservoir development is a block/cluster water injection with rows of production wells having 300 × 300 m spacing. The depth is 1600 m, formation pressure about 15 MPa, temperature of 75 °C, oil viscosity of 2 mPa s, and oil density of 820 kg m⁻¹.

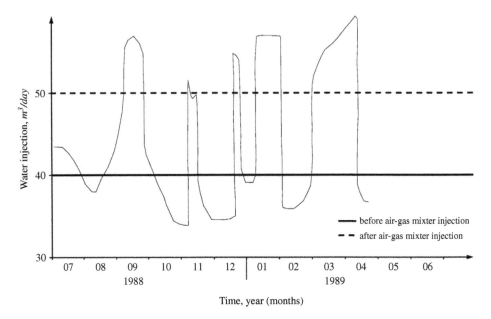

Figure 13.6 Water injection dynamics.

Field trials have been carried out in the pattern area containing six injection wells and 26 production wells. The daily water injection rate has been maintained at the same level as it was in a water-flooding case (e.g. 200–300 m^3/well). A decrease by 10–15% in the water production rate was observed and a 15–20% increase in oil production compared to the water-flooding case.

Nomenclature

a_1 and a_2	rational numbers
η_w	water viscosity, Pa s
η_1, η_2	apparent oil viscosity taking account of the slippage for the high- and low permeability layers, Pa s
η_0	real oil viscosity, Pa s

References

1 Chen, Y., Wang, Y., Lu, J., and Wu, C. (2009). The viscosity reduction of nano-keggin-K3PMo12O40 in catalytic aquathermolysis of heavy oil. *Fuel* 88 (8): 1426–1434.
2 Ogolo, N.A., Olafuyi, O.A. and, Onyekonwu M.O. Enhanced oil recovery using nanoparticles. *SPE-160847-MS. Presented at the SPE Saudi Arabia Section Technical Symposium and Exhibition*, Al-Khobar, Saudi Arabia (8–11 April 2012).
3 Ehtesabi, H., Ahadian, M.M., and Taghikhani, V. (2015). Enhanced heavy oil recovery using TiO_2 nanoparticles: investigation of deposition during transport in core plug. *Energy & Fuels* 29 (1): 1–8.
4 Alnarabiji, M.S., Yahya, N., Nadeem, S. et al. (2018). Nanofluid enhanced oil recovery using induced ZnO nanocrystals by electromagnetic energy: viscosity increment. *Fuel* 233: 632–643.
5 Sun, Q., Li, Z., Li, S. et al. (2014). Utilization of surfactant-stabilized foam for enhanced oil recovery by adding nanoparticles. *Energy & Fuels* 28 (4): 2384–2394.
6 Maghzi, A., Mohebbi, A., Kharrat, R., and Ghazanfari, M.H. (2011). Pore-scale monitoring of wettability alteration by silica nanoparticles during polymer flooding to heavy oil in a five-spot glass micromodel. *Transport in Porous Media* 87 (3): 653–664.
7 Bayat, A.E., Junin, R., Shamshirband, S., and Chong, W.T. (2015). Transport and retention of engineered Al_2O_3, TiO_2, and SiO_2 nanoparticles through various sedimentary rocks. *Scientific Reports* 5: 14264.
8 Alomair, O.A., Matar, K.M. and, Alsaeed Y.H. Nanofluids application for heavy oil recovery. *SPE-171539-MS. Presented at the SPE Asia Pacific Oil & Gas Conference and Exhibition*, Adelaide, Australia (14–16 October 2014).
9 Sharma, T., Kumar, G.S., and Sangwai, J.S. (2015). Comparative effectiveness of production performance of Pickering emulsion stabilized by nanoparticle–surfactant–polymerover surfactant–polymer (SP) flooding for enhanced oil recoveryfor Brownfield reservoir. *Journal of Petroleum Science and Engineering* 129: 221–232.
10 Suleimanov, B.A., Ismailov, F.S., and Veliyev, E.F. (2011). Nanofluid for enhanced oil recovery. *Journal of Petroleum Science and Engineering* 78 (2): 431–437.
11 Shamsi, J.H., Miller, C.A., Wong, M.S. et al. (2014). Polymer-coated nanoparticles for enhanced oil recovery. *Journal of Applied Polymer Science* 131 (15).
12 Pei, H., Zhang, G., Ge, J. et al. (2015). Investigation of synergy between nanoparticle and surfactant in stabilizing oil-in-water emulsions for improved heavy oil recovery. *Colloids and Surfaces A: Physicochemical and Engineering Aspects* 484: 478–484.

Nanocolloids for Petroleum Engineering: Fundamentals and Practices, First Edition.
Baghir A. Suleimanov, Elchin F. Veliyev, and Vladimir Vishnyakov.
© 2022 John Wiley & Sons Ltd. Published 2022 by John Wiley & Sons Ltd.

13 Hendraningrat, L. and Torsæter, O. (2016). A study of water chemistry extends the benefits of using silica-based nanoparticles on enhanced oil recovery. *Applied Nanoscience* 6 (1): 83–95.

14 Hendraningrat, L., Li, S., and Torsæter, O. (2013). A coreflood investigation of nanofluid enhanced oil recovery. *Journal of Petroleum Science and Engineering* 111: 128–138.

15 Hendraningrat, L. and Torsæter, O. (2014). Effects of the initial rock wettability on silica-based nanofluid-enhanced oil recovery processes at reservoir temperatures. *Energy & Fuels* 28 (10): 6228–6241.

16 Amanullah, M. and Al-Tahini, A.M. (2009). Nano-technology – Its significance in smart fluid development for oil and gas field application. *SPE-126102-MS. Presented at the SPE Saudi Arabia Section Technical Symposium*, Al-Khobar, Saudi Arabia (9–11 May 2009).

17 Zitha, P.L.J. (2005). Smart fluids in the oilfield. In: *Exploration & Production: The Oil & Gas Review*, 66–68.

18 Romanovsky, B.V. and Makshina, E.V. (2004). Nanocomposites as functional materials. *Soros Educational Journal* 8 (2).

19 Kostic, M.M. and Cho,i S.U.S. Cristial issues and application potentials in nanofluids research. *Proceedings of MN2006 Multifunctional Nanocomposites*, Honolulu, Hawaii (20–22 January 2006).

20 Suleimanov, B.A. (2006). Specific features of heterogeneous systems flow in porous media. Moscow-Izhevsk, Institute of Computer Science (Series: *Modern Oil and Gas Technologies*).

21 Wasan, D.T. and Nikolov, A.D. (2003). Spreading of nanofluids on solids. *Nature* 423 (6936): 156–159.

22 Munshi, A.M., Singh, V.N., Kumar, M., and Singha, J.P. (2008). Effect of nanoparticle size on sessile droplet contact angle. *Journal of Applied Physics* 103: 084315.

23 Ravera, F., Santini, E., Loglio, G. et al. (2006). Effect of nanoparticles on the interfacial properties of liquid/liquid and liquid/air surface layers. *Journal of Physical Chemistry B* 110 (39): 19543–19551.

24 Suleimanov, B.A., Veliyev, E.F., and Aliyev, A.A. (2020). Colloidal dispersion nanogels for in-situ fluid diversion. *Journal of Petroleum Science and Engineering* 193: 107411.

25 Veliyev, E.F. (2020). Review of modern in-situ fluid diversion technologies. *SOCAR Proceedings* 2: 50–66.

26 Sydansk, R. (1987). Conformance improvement in a subterranean hydrocarbon-bearing formation using a polymer gel. Patent US 4683949 A.

27 Singh, P., Venkatesan, R., Fogler, H.S., and Nagarajan, N. (2000). Formation and aging of incipient thin film wax-oil gels. *AIChE Journal* 46: 1059–1074.

28 Meister, J. Bulk gel strength tester. *SPE-13567-MS. Presented at the SPE Oilfield and Geothermal Chemistry Symposium*, Phoenix, Arizona (April 1985).

29 Romeo-Zeron, L. (2008). Characterization of crosslinked gel kinetics and gel strength by use of NMR. *SPE Reservoir Evaluation & Engineering* 11: 3–18.

30 Norisuye, T., Strybulevych, A., Scanlon, M., and Page, J. (2006). Ultrasonic investigation of the gelation process of poly (acrylamide) gels. *Macromolecular Symposia* 242 (1): 208–215.

31 Blitz, J. and Simpson, G. (1996). *Ultrasonic Methods of Non-destructive Testing*. Champman & Hall.

32 Hanemann, T. and Szabó, D.V. (2010). Polymer-nanoparticle composites: from synthesis to modern applications. *Materials* 3 (6): 3468–3517.

33 Gerard, J.F. (2001). *Fillers and Filled Polymers*. Weinheim: Wiley-VCH.

34 Postan, M.Y. (1993). Generalized logistic curve: properties and estimate of parameters. *Economics and Mathematical Methods* 29 (2): 305–310.

35 Mamunya, Y.P., Davydenko, V.V., Pissis, P., and Lebedev, E.V. (2002). Electrical and thermal conductivity of polymers filled with metal powders. *European Polymer Journal* 38: 1887–1897.

36 Sharma, P., Miao, W., Giri, A., and Raghunathan, S. (2004). Nanomaterials: Manufacturing, processing, and applications. In: *Dekker Encyclopedia of Nanoscience and Nanotechnology*.

37 Stauer, D. and Aharony, A. (1994). *Introduction to Percolation Theory*. London: Taylor & Francis.

38 Adler, J. (1991). Bootstrap percolation. *Physica A* 171: 453.

39 Jensen, I. (2000). Enumerations of lattice animals and trees. *Journal of Statistical Physics* 102: 865.

40 American Petroleum Institute (1998). *API RP40. Recommended Practices for Core Analysis*, 2e. Washington, DC: API.

41 Entov, V.M. and Turetskaya, F.D. (1995). Hydrodynamical modeling of the development of non-homogeneous oil reservoirs. *Fluid Dynamics* 30: 877–882.

42 Song, Z., Liu, L., Wei, M. et al. (2015). Effect of polymer on disproportionate permeability reduction to gas and water for fractured shales. *Fuel* 143: 28–37.

43 Suleimanov, B.A. and Veliyev, E.F. (2017). Novel polymeric nanogel as diversion agent for enhanced oil recovery. *Petroleum Science and Technology* 35 (4): 319–326.

44 Suleimanov, B.A., Veliyev, E.F. Nanogels for deep reservoir conformance control. *SPE-182534-MS. Presented at the SPE Annual Caspian Technical Conference & Exhibition*, Astana, Kazakhstan(1–3 November 2016).

45 Ayirala, S., Doe, P., Curole, M., and Chin, R. (2008). Polymer flooding in saline heavy oil environments. *Paper SPE-113396. Presented at the SPE/DOE Symposium on Improved Oil Recovery*.

46 Suleimanov, B.A., Veliyev, E.F., and Dyshin, O.A. (2015). Effect of nanoparticles on the compressive strength of polymer gels used for enhanced oil recovery (EOR). *Petroleum Science and Technology* 33 (10): 1133–1140.

47 Suleimanov, B.A., Dyshin, O.A., and Veliyev, E.F. Compressive strength of polymer nanogels used for enhanced oil recovery EOR. *SPE-181960-MS. Presented at the SPE Russian Petroleum Technology Conference and Exhibition*, Moscow, Russia (24–26 October 2016).

48 Spildo, K., Skauge, A., Aarra, M.G., and Tweheyo, M.T. (2009). A new polymer application for North Sea reservoirs. *SPE Reservoir Evaluation & Engineering* 12 (3): 427–432.

49 Bjorsvik, M., Hoiland, H., and Skauge, A. (2007). Formation of colloidal dispersion gels from aqueous polyacrylamide solutions. *Colloids and Surfaces A: Physicochemical and Engineering Aspects* 317: 504–511.

50 Al-Assi, A.A., Willhite, G.P., Green, D.W., and McCool, C.S. (2009). Formation and propagation of gel aggregates using partially hydrolyzed polyacrylamide and aluminum citrate. *SPE Journal* 14: 450–461.

51 Frampton H., Morgan J.C., Cheung S.K., et al. Development of a novel waterflood conformance control system. *SPE-89391-MS. Presented at the SPE/DOE Symposium on Improved Oil Recovery*, Tulsa (17–21 April 2004).

52 Breulmann, M., Davis, S.A., Mann, S. et al. (2000). Polymer-gel templating of porous inorganic macro-structures using nanoparticle building blocks. *Advanced Materials* 12 (7): 502–507.

References

53 Suleimanov, B.A. and Veliyev, E.F. (2016). The effect of particle size distribution and the nano-sized additives on the quality of annulus isolation in well cementing. *SOCAR Proceedings* 4: 4–10.

54 Suleimanov, B.A., Ismailov, F.S., Veliyev, E.F., and Dyshin, O.A. (2013). The influence of light metal nanoparticles on the strength of polymer gels used in oil industry. *SOCAR Proceedings* 2: 24–28.

55 Ranganathan, R., Lewis, R., McCool, C.S. et al. (1998). Experimental study of the gelation behavior of a polyacrylamide/aluminum citrate colloidal-dispersion gel system. *SPE Journal* 3 (4): 337–343.

56 Spildo, K., Skauge, A., and Skauge, T. Propagation of colloidal dispersion gels (CDG) in laboratory corefloods. *SPE-129927-MS. Presented at the SPE Improved Oil Recovery Symposium*, Tulsa, Oklahoma, USA (24–28 April 2010).

57 Mack, J.C. and Smith, J.E. In-depth colloidal dispersion gels improve oil recovery efficiency. *Paper SPE-27780-MS. Presented at the SPE/DOE Ninth Symposium on Improved Oil Recovery*, Tulsa, Oklahoma (April 17–20 1994).

58 Diaz, D., Somaruga, C., Norman, C. and Romero, J. Colloidal dispersion gels improve oil recovery in a heterogeneous Argentina waterflood. *SPE-113320-MS. Presented at the SPE Symposium on Improved Oil Recovery*, Tulsa, Oklahoma, USA (20–23 April 2008).

59 Skauge, T., Spildo, K., Skauge, A. and (2010). Nano-sized particles for EOR. *SPE-129933-MS. Presented at the SPE Improved Oil Recovery Symposium*, Tulsa, Oklahoma, USA (24–28 April 2010).

60 Fielding, R.C. Jr., Gibbons, D.H., and Legrand, F.P. In-depth drive fluid diversion using an evolution of colloidal dispersion gels and new bulk gels: an operational case history of North Rainbow Ranch unit. *SPE-27773-MS. Presented at the SPE/DOE Improved Oil Recovery Symposium*, Tulsa, Oklahoma (17–20 April 1994).

61 Veliyev, E.F., Aliyev, A.A., and Guliyev, V.V., Naghiyeva, N.V. (2019). Water shutoff using crosslinked polymer gels. *SPE-198351-MS. Presented at the SPE Annual Caspian Technical Conference*, Baku, Azerbaijan (16–18 October 2019).

62 Smith J.E. Performance of 18 polymers in aluminum citrate colloidal dispersion gels. *SPE-28989-MS. Presented at the SPE International Symposium on Oilfield Chemistry*, San Antonio, Texas (14–17 February 1995).

63 Spildo, K., Skauge, A., Aarra, M.G., and Tweheyo, M.T. (2009). A new polymer application for North Sea reservoirs (SPE-113460-PA). *SPE Reservoir Evaluation & Engineering* 12 (3): 427–432.

64 Schexnailder, P. and Schmidt, G. (2009). Nanocomposite polymer hydrogels. *Colloid and Polymer Science* 287 (1): 1–11.

65 Liu, F. and Urban, M.W. (2010). Recent advances and challenges in designing stimuli-responsive polymers. *Progress in Polymer Science* 35 (1–2): 3–23.

66 Natarajan D., McCool C.S., and Green D.W., Willhite G. P. Control of in-situ gelation time for HPAAM-chromium acetate systems. *SPE-39696-MS. Presented at the SPE/DOE Improved Oil Recovery Symposium*, Tulsa, Oklahoma (19–22 April 1998).

67 Wang, D., Han, P., Shao, Z. et al. (2008). Sweep improvement options for the Daqing oil field. *SPE Reservoir Evaluation & Engineering* 11 (1): 18–26.

68 Broseta, D., Marquer, O., Blin, N., and Zaitoun, A. (2000). Rheological screening of low-molecular-weight polyacrylamide/chromium(III) acetate water shutoff gels. *SPE-59319-MS*.

Presented at the SPE/DOE Improved Oil Recovery Symposium, Tulsa, Oklahoma (3–5 April 2000).

69 Vishnyakov, V., Suleimanov, B., Salmanov, A., and Zeynalov, E. (2019). *Primer on Enhanced Oil Recovery*. Gulf Professional Publishing.

70 Coste J.P., Liu Y., Bai B., et al. In-depth fluid diversion by pre-gelled particles. Laboratory study and pilot testing. SPE-59362-MS. *Presented at the SPE/DOE Improved Oil Recovery Symposium*, Tulsa, Oklahoma (3–5 April 2000).

71 Murshed, S.M.S., Tan, S.-H., and Nguyen, N.-T. (2008). Temperature dependence of interfacial properties and viscosity of nanofluids for droplet-based microfluidics. *Journal of Physics D: Applied Physics* 41 (8): 085502.

72 Suleimanov, B.A., Ismayilov, R.H., Abbasov, H.F. et al. (2017). Thermophysical properties of nano-and microfluids with [Ni5 (µ5-pppmda) 4Cl$_2$] metal string complex particles. *Colloids and Surfaces A: Physicochemical and Engineering Aspects* 513: 41–50.

73 O'Brien, R.W. (1990). Electroacoustic studies of moderately concentrated colloidal suspensions. *Faraday Disc. Chemical Society* 90: 301–312.

74 Hanaor, D., Michelazzi, M., Leonelli, C., and Sorrell, C.C. (2012). The effects of carboxylic acids on the aqueous dispersion and electrophoretic deposition of ZrO_2. *Journal of the European Ceramic Society* 32: 235–244.

75 Mirzajanzade, A. H. and Ametov, I. M. Rheotechnology – A new trend in oil production technology. In *Proceedings of the 6th European Symposium on Improved Oil Recovery*, Stavanger (May 1991).

76 Mirzadjanzade, A.Kh., Ametov, I.M., and Bogopolsky, A.O., et al. Micronucleous water-gas solutions: A new efficient agent for water-flooded oil reservoirs. *Proceedings of the 7th European IOR Symposium*, Moscow, Russia (October 1993).

77 Suleimanov, B.A., Azizov, K.F., and Abbasov, E.M. (1996). Slippage effect during gassed oil displacement. *Energy Sources* 18 (7): 773–779.

78 Suleimanov, B.A. (1997). Slip effect during flow of gassed liquid. *Colloid Journal* 59 (6): 749–753.

79 Suleimanov, B.A., Azizov, K.F., and Abbasov, E.M. (1998). Specific features of the gas-liquid mixture filtration. *Acta Mechanica* 130 (1–2): 121–133.

Part E

Novel Perspective Nanocolloids

14

Metal String Complex Micro and Nano Fluids

14.1 What are Metal String Complexes?

Molecular electronic devices that formed from a packet of molecules could be a good alternative for conventional silicon-based semiconductors [1, 2]. The small sizes (e.g. up to a hundred times smaller than the conventional semiconductors), a cost effective manufacturing process, and large property variations of the final products are strong arguments for predicting a perspective future development. Molecular chains that conduct an electrical current (i.e. molecular wires) are the building blocks of molecular electronic devices. Among a large variety of molecular wires, researchers have particularly focused on branch complexes with one-dimensional linear transition metal backbones that are expected to transfer electrons similarly to traditional electric wires. The "Krogmann salt" was the first composite used [3]. Peng et al. successfully designed a metal string complex (MSC) supported by four ligands (e.g. $Ni_3(dpa)_4Cl_2$ (dpa- = dipyridylamido anion)) in 1968, but the structure was determined only after 20 years [4]. The novelty of the obtained geometry encouraged researchers to synthesize and study similar compounds containing metal atoms (where the metals can be Co, Cr, Ru, and Rh) [4]. Meantime, Professor F.A. Cotton has made a big contribution by providing significant insights into the physical properties of MSCs [5].

It should be noted that there are two different schools of thought with different terminology to name the obtained compounds. Thus, Peng's school call it "Metal String Complexes" whilst Cotton used "Extended Metal Atom Chains (EMACs)" [5].

MSCs have a large variety of metal–metal bonds due to the existence of extra metal ions and paddlewheel geometry. The variety is the main reason for researchers to concentrate on the study of multinuclear multiple metal–metal bonded compounds. Berry provided a detailed review of the electronic structure and bonding mechanism of MSCs [5].

The first studies were mainly dedicated to the trinuclear MSCs. However, it was noted that an increase of the transition metal ions was leading to different multinuclear metal–metal bonds. Unfortunately, the early studies were limited by the then available computational capacities, which did not allow properly obtained data analysis. MSCs were studied in terms of electronic transport by electrochemical and voltammetric methods, which demonstrate useful characteristics of molecular wire [6, 7]. In the case of strong metal–metal bonds, the conductivity significantly increases [7, 8].

Nanocolloids for Petroleum Engineering: Fundamentals and Practices, First Edition.
Baghir A. Suleimanov, Elchin F. Veliyev, and Vladimir Vishnyakov.
© 2022 John Wiley & Sons Ltd. Published 2022 by John Wiley & Sons Ltd.

In the last decade Peng et al. have presented two different branches of new generation MSCs:

- Mixed-valence oligonickel string complexes supported by various naphthyridyl-modulated ligands [9].
- Heteronuclear metal string complexes (HMSCs) based on inserting a heterometal ion into the homonuclear MSCs.

Apart from high conductivity values, the mentioned compounds also demonstrate an excellent stability that makes them the most perspective MSCs currently studied. The lower density of Ni MSCs compared to microparticles is the biggest achievement for nanogel high stability (e.g. 3.5 times lower than pure Ni) [10, 11]. This chapter will analyze the impact of the presence of MSCs on IFT, rheology, stability, and thermal conductivity properties of the fluid. The results obtained demonstrate a promising prospect of MSCs for utilization in enhanced oil recovery.

14.2 Thermal Conductivity Enhancement of Microfluids with $Ni_3(\mu 3\text{-ppza})_4Cl_2$ Metal String Complex Particles

A suspension of solid particles in liquid has some unique physical properties that differ from those of the base fluid. Maxwell first raised the idea of the possibility of improving the physical properties of a fluid, in particular its thermal conductivity, by adding solid particles [12]. Then, in 1995 at the Argonne National Laboratory, significant thermal-conductivity enhancement was first detected after adding nanoparticles to a fluid [13].

A nanofluid (microfluid) is a colloid obtained by dispersing nanoparticles (microparticles) in a conventional fluid; it is a two-phase system consisting of a base fluid and solid particles. The base fluid can be water, an organic liquid (e.g. ethylene glycol, glycerol, oil), or a solution of the two (e.g. ethylene-glycol/water, glycerol/water) [14–16]. The nanoparticles can be particles of chemically stable metals and their oxides (e.g. Cu, CuO, Al, Al_2O_3, Zn, Fe, Au, Ag, etc.), carbon nanotubes, or polymeric nanoparticles, and the microparticles can be certain MSC particles [11].

Among many possibilities, nanofluids and microfluids have the potential to be used (i) in electronics for cooling various microelectronic devices, (ii) in the power industry for rapid and efficient heat transport from heat sources (e.g. in nuclear reactors), (iii) in the oil industry for increasing oil production, (iv) for creating new medicines, cosmetics, lubricants, varnishes, and paints, and (v) for delivering medicines [15–19]. Nanofluids and microfluids can also be used successfully in engineering (e.g. as antifreeze [20]) and in everyday life as better heat transfer media in various devices.

An analysis of the literature indicates that despite the plurality of different models [21–30] for heat transfer in nanofluids, there are many disagreements between existing theories and experiments, and a clear understanding of the main mechanisms involved in higher thermal conductivity of nanofluids has not yet been established.

When using suspensions containing micron-size particles of ordinary metals and their oxides, Maxwell's predicted increase in the thermal conductivity of liquids had not been observed before due to the fact that normally the particles rapidly sediment. However, it was shown later that microfluids can have a high thermal conductivity (comparable to that of nanofluids) [11] if microparticles of the metal complex Ni_5 are used. The fact that the effect can be observed in this case is attributed to the high stability of microfluids due to the formation of pseudo-periodic structures in the colloidal solution, and also to the fact that the particles of the MSC used are single-crystal ones. A mathematical model based on experimental data was used to explain the enhancement of thermal conductivity and the effect of particle aggregation on the thermal properties of nanofluids and microfluids [31].

14.2.1 Microparticles of MSC $Ni_3(\mu3\text{-ppza})_4Cl_2$

Microparticles were synthesized for the first time by reacting anhydrous $NiCl_2$ with N-(pyridin-2-yl)pyrazin-2-amine ligand (Hppza) under an argon atmosphere. Naphthalene was used as the solvent and also as a base to deprotonate the amine group. This process was conducted according to the previously reported method for other trinickel string complexes, as described below [32–34]. To obtain $Ni_3(\mu3\text{-ppza})_4Cl_2$, anhydrous $NiCl_2$ (0.69 g, 5.3 mmol), Hppza (1.00 g, 5.81 mmol), and naphthalene (30 g) were placed in an Erlenmeyer flask. The mixture was heated at 170–180 °C for 12 hours under argon, after which a solution of potassium tert-butoxide (0.71 g, 6.35 mmol) in n-butyl alcohol (4 ml) was added dropwise. The mixture was refluxed continuously for 6 hours. After the mixture had cooled, hexane was added to wash out the naphthalene and then 200 ml of CH_2Cl_2 was used to extract MSC. Dark-brown rod-shaped single crystals suitable for X-ray diffraction were obtained from diffusing ether to a dichloromethane solution (0.32 g, 23.7% yield). An X-ray crystal-structure analysis revealed that, similar to other oligo-α-pyridylamino MSCs, the tri-nickel linear metal chain in $Ni_3(\mu3\text{-ppza})_4Cl_2$ is wrapped helically by four syn–syntypeppza ligands. All of the nickel ions and two axial ligands are nearly collinear, with the Ni–Ni–Ni angle equal to 179.35°. The whole Ni_3 chain is roughly 4.87 Å long (see Figure 14.1a) and the bulk density is at around 1.644 g cm^{-3}.

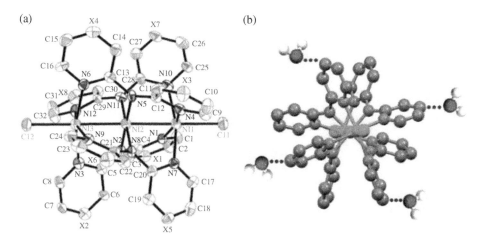

Figure 14.1 (a) Crystal structure of $Ni_3(\mu_3\text{-ppza})_4Cl_2$ (where X = 1/2 N + 1/2 C). The thermal ellipsoids are at the 30% probability level and hydrogen atoms are omitted for clarity. (b) Hydrogen bonds (dashed lines) between metal string complex (MSC) particles (through their organic fragments) and water molecules.

To conduct experiments, a sufficient amount of a mixture of 70 wt% distilled water and 30 wt% glycerol was prepared as the base fluid. The possibility of applying this mixture as a coolant in automobile radiators [17, 18] was the main reason for choosing this mixture as the base fluid.

14.2.2 Ni$_3$ Microfluid

Ni$_3$ microparticles were dispersed in the base fluid. Monocrystalline MSC [Ni$_3$(μ_5-pppmda)$_4$Cl$_2$] microparticles were obtained by a mechanical ball milling method. DLS data show that the size of the monocrystalline MSC [Ni$_3$(μ_5-pppmda)$_4$Cl$_2$] microparticles in the base fluid is around 1 µm and they are monodispersed in size, with a polydispersity of around 8% (Figure 14.2).

Monocrystalline Ni$_3$(μ_3-ppza)$_4$Cl$_2$ microparticles were processed utilizing a mechanical ball mill (Planetar PM 100; Retsch, Germany). In addition to the well-proven mixing and size-reduction processes, this mill also meets all of the technical requirements for colloidal grinding and provides the necessary energy input for mechanical alloying.

14.2.3 Fluid Stability

The particle size distribution of each microfluid was measured using a dynamic light-scattering (DLS) particle-size distribution analyzer (LB-550; Horiba, Japan). The morphologies and sizes of the microparticles were studied using a scanning electron microscope (SEM) (JSM-7600F; JEOL, Japan).

Monocrystalline MSC Ni$_3$(μ_3-ppza)$_4$Cl$_2$ particles were well distributed in the base fluid and needed no additional treatments to stabilize them. There are at least three reasons for the better stabilization of the Ni$_3$ complex microparticles in the base fluid:

– *Hydrogen bonds.* The hydrogen bonds are formed between the MSC particles (through their organic fragments) and water molecules (see Figure 14.1b).

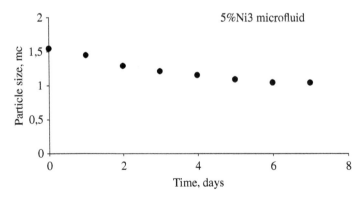

Figure 14.2 Stabilization of particle size of the monocrystalline metal string Ni$_3$(μ_3-ppza)$_4$Cl$_2$ complex (5 vol%) in base fluid.

- *Density.* The density of Ni_3 complex microparticles ($\rho_m = 1.644$ g cm^{-3}) is much smaller compared to Cu nanoparticles ($\rho_{Cu} = 5$ g cm^{-3}) and is comparable with Ni_5 complex microparticles ($\rho_m = 1.668$ g cm^{-3}), though greater than the base fluid ($\rho_{bf} = 1.07$ g cm^{-3}).
- *Association.* The SEM micrographs (Figure 14.3) clearly show the formation of microparticle assemblies whose dimensions are determined by the equilibrium between gravity and buoyancy.

Figure 14.3 shows that the average radius of the assembly (R_a) is 2.5 μm and the thickness of the associated surrounding liquid layer (H) is approximately 0.2 μm, the average radius of the microparticles (r_p) can be taken equal to 0.7 μm, and the average number of particles in the assembly (N) is approximately 14. According to reference [35], at large interparticle separations (compared to atoms), such as the 0.2 μm observed in our measurements, stable "secondary minimum" associations between colloidal particles can form with high degrees of order and with a thixotropic colloidal structure.

The value of microparticle density at which microfluid stabilization is achieved was determined from the condition of buoyancy, when the force of gravity acting on the assembly is balanced by the buoyancy:

$$N \frac{4}{3}\pi (r_p + H)^3 \rho_p g = \frac{4}{3}\pi R_a^3 \rho_{bf} g$$

where g is the acceleration due to gravity. Substituting in the formula the values of the parameters, we get $\rho_p = 1.638$ g cm^{-3}, which is practically the same as the density of Ni_3 complex particles.

Figure 14.3 Scanning electron microscope images of microparticles of MSC $Ni_3(\mu_3\text{-ppza})_4Cl_2$ (1 vol%) in base fluid.

14.2.4 Thermal Conductivity

The transient hot-wire method involves measuring how the temperature increases over time when current is passed through a thin electrically insulated wire immersed in the test liquid [36, 37]. The investigated liquid was maintained in a thermostat at a predetermined temperature. The heated wire acts as a heat source, from which a thermal wave spreads inside a homogeneous fluid. The theoretical model describing this method is based on solving the one-dimensional heat equation in cylindrical coordinates for an infinitely long linear heat source of negligible radius. In this model, the temperature rise at the radial distance r from the heat source is

$$\Delta T(r,t) = T(r,t) - T_0 = \frac{q}{4\pi k} \ln\left(\frac{4at}{r^2 C}\right) \tag{14.1}$$

where T_0 is the initial temperature, q is the thermal flux per unit wire length, k is the thermal conductivity coefficient of the homogeneous fluid through which the heat propagates, $a = k/c\rho$ is the thermal diffusivity, c is the specific heat of fluid, ρ is the fluid density, and $C = \exp(\gamma)$, where $\gamma = 0.5772157$ is Euler's constant.

Equation (14.1) shows that by constructing a linear-dependence graph $T(\ln t)$, we can find the thermal conductivity coefficient $k = q/(4s)$ via the slope (s) of the straight section. Alternatively, by comparing the graph slopes for investigated (k_f) and reference (k_{et}) fluids, the thermal conductivity coefficient of the tested fluid is determined as $k_f = k_{et} s_{et}/s_f$.

The thermal conductivities of the studied dispersions were measured on a special setup (see Figure 14.4), a central element of which is a platinum wire of length $L = 95$ mm and resistance $R \approx 0.4\ \Omega$. For the setup and tested fluids, the accepted timescales related to the flow and heat transfer were $t_{min} \geq 0.7c$ and $t_{max} \leq 240c$. All measurements were performed at a constant heat output per unit wire length of $I^2 R/L \approx 4.21$ W m^{-1} with a current of $I = 1$ A

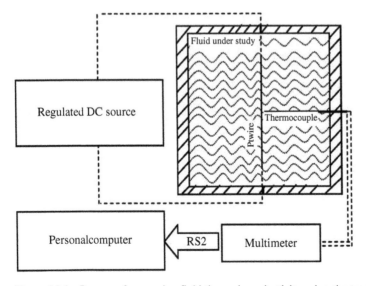

Figure 14.4 Process of measuring fluid thermal conductivity using the transient hot-wire method.

14.2 Thermal Conductivity Enhancement of Microfluids with $Ni_3(\mu 3\text{-}ppza)_4Cl_2$ Metal String Complex Particles

at room temperature. A DC power supply (PS-3005D, NMX Technology Co., Ltd., Hong Kong) was used as a stabilized source. The time dependence of the temperature was measured by transmitting a signal from a thermocouple end, which closely leaned to the middle of a platinum wire covered by a thin layer of varnish, through a multimeter (MT-1860, Prokit's Industries Co., Ltd., Taiwan) and to a computer with an RS232 interface. The data received by the computer were processed and a graph of the temperature versus log time, $T(\ln t)$, was constructed for each sample. The thermal conductivity $k_f = k_{et} S_{et}/S_f$ of each fluid was determined by comparing the slope s of the straight-line portion of its graph with that for the reference fluid, which was distilled water at room temperature with $k_{et} = 0.6$ W(mK)$^{-1}$.

Figure 14.5 shows the thermal-conductivity enhancement for Ni_3–water–glycerol as a function of the volume fraction of microparticles. Figure 14.6 shows previously obtained values of thermal-conductivity enhancement for certain nanofluids and microfluids at a particle volume fraction of 5 vol% [11, 31]. A comparison shows that the best results were achieved using microparticles of monocrystalline MSCs $Ni_3(\mu_3\text{-}ppza)_4Cl_2$ as well as $Ni_5(\mu_5\text{-}pppmda)_4Cl_2$ and copper nanoparticles. Compared to the base fluid, the thermal conductivity coefficient increased by 72% for Ni_3–water–glycerol, 53% for Cu–water–glycerol, and 47% for Ni_5–water–glycerol. The results were lower for Al nanofluid and Ni microfluids: roughly 20% for 60 nm Al nanofluid, 10% for 100 nm Al nanofluid, and only around 5% for Ni microfluid. The results can be explained within the framework of the proposed model of thermal-conductivity enhancement in nanofluids (microfluids) [31]:

$$\frac{k_{ef} - k_f}{k_f} = \frac{k_n}{k_f} \frac{\phi r_f \left[1 + c\left(\frac{r_{cl}}{r_n}\right)^2\right]}{(1-\phi) r_n \left[1 + c\left(\frac{r_{cl}}{r_n}\right)^3\right]} \tag{14.2}$$

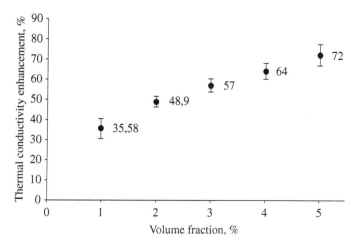

Figure 14.5 Concentration dependence of thermal-conductivity enhancement for Ni_3–water–glycerol.

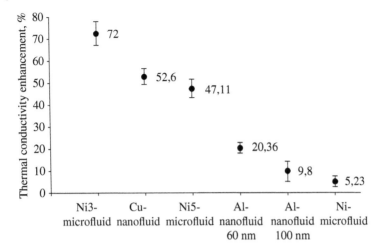

Figure 14.6 Thermal-conductivity enhancement of nano- and microfluids at a particle volume fraction of φ = 5 vol%.

where k_{ef} is the dispersion thermal conductivity coefficient, k_f is the base-fluid thermal conductivity coefficient, k_n is the particle thermal conductivity coefficient, φ is the particle volume fraction, r_n is the particle radius, r_{cl} is the radius of a cluster formed by particle aggregation, and r_f is the radius of a base fluid molecule.

It can be seen from Equation (14.2) that the thermal-conductivity enhancement of dispersions with low concentrations of nanoparticles is directly proportional to the volume fraction (see Figure 14.5) and the coefficient of thermal conductivity (again, see Figure 14.5) and inversely proportional to the particle radius (see Figure 14.6) and the particle aggregation coefficient c.

The advantage of using microparticles instead of nanoparticles with the view to increase the thermal conductivity is that microparticles are easy to obtain by simple mechanical crushing in laboratory mills without the high additional costs that are necessarily incurred in obtaining nanoparticles. It was suggested that high thermal-conductivity enhancement achieved with Ni_3 MSC microparticles is due to the colloid stability, which is achieved by the formation of hydrogen bonds between the MSC particles (through their organic fragments) and water molecules (see Figure 14.1b). The absence of such bonds between microparticles of ordinary metals (and their oxides) and liquid molecules causes them to sediment rapidly in the liquid, thereby preventing any increased thermal conductivity being observed upon adding such microparticles to the liquid. The increased thermal conductivity of a microfluid of Ni_5 MSC microparticles is also associated with the formation of hydrogen bonds between the MSC particles (again through their organic fragments) and liquid molecules. The higher thermal-conductivity enhancement for Ni_3–water–glycerol (72% at a volume fraction of 5 vol%) compared to Ni_5–water–glycerol (50% at a volume fraction of 5 vol%) is most likely due to the higher thermal conductivity of Ni_3 MSC compared to Ni_5 MSC. This can be rationalized as follows. It is known that the main mechanism for the thermal conductivity of metals is electronic, which is also responsible for their electrical conductivity.

According to the Wiedemann–Franz law, the ratio of the electronic contribution of the thermal conductivity to the electrical conductivity of a metal is proportional to the temperature. It has been established that the electrical conductivity of Ni_3 MSC is higher compared to Ni_5 MSC [38].

14.2.5 Rheology

The rheology of the investigated systems was studied using the Brookfield Model BEL-PVSR-230 Rheometer (USA), which was designed to test small complex samples by simulating process conditions in a bench top environment and can be also used for measuring the temperature dependence of fluid viscosity.

Measuring the viscosities of the studied suspensions showed them to be slightly higher than the base-fluid viscosity (Figure 14.7) (accuracy in viscosity measurement is 0.1 mPa.s). However, it is obvious that the base fluid is Newtonian, but nanoparticle addition changes its rheological behavior to that of a pseudo-plastic. At low shear rates the nanoparticles, in addition to translational motion, are able to rotate, a fact that significantly contributes to the viscosity enhancement of nanofluids. The behavior of the viscosity in the case of the associated complex is determined by the thixotropic colloidal structure. In such systems the increase in shear rate destroys the colloidal structure, resulting in viscosity reduction [39, 40]. A comparison of rheological behavior shows that the viscous friction force in the fluid flow will be lower in the case of Ni_3–water–glycerol than of Ni_5–water–glycerol [11].

14.2.6 Surface Tension

The surface tension of each investigated microfluid was measured using a tensiometer (DSA30; Krüss Scientific, Germany) using the pendant-drop method.

The concentration dependence of surface tension (γ_l) for Ni_3–water–glycerol is shown in Figure 14.8 (accuracy in the surface tension measurement is 0.1 mN m^{-1}). The surface tension decreases slightly as the concentration is increased from zero but then becomes

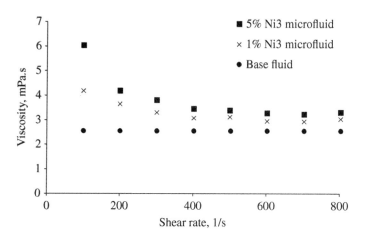

Figure 14.7 The dependence of viscosity on shear rate for Ni_3-microfluid at a temperature of 25 °C.

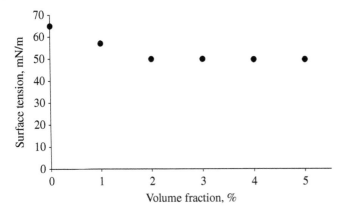

Figure 14.8 Concentration dependence of surface tension of Ni_3-water-glycerol.

constant for concentrations in excess of 2 vol%. This means that the wetting (S) of this dispersion is lower compared to that of the base fluid, and the suspension will flow with less hydrodynamic resistance: $S = \gamma_S + \gamma_l - \gamma_{Sl}$, where γ_S is the surface tension of the solid and γ_{Sl} is the interfacial tension between the solid and the liquid.

14.2.7 Freezing Points

The concentration dependence of the freezing point of Ni_3-water-glycerol is shown in Figure 14.9 (accuracy in the freezing point measurement is 0.1 °C). An increase in the concentration of Ni_3 MSC particles leads to a drop in the freezing point, which is a basic requirement for any liquid used as antifreeze in a machine radiator for cooling. A comparison of the freezing points of microfluids and nanofluids in the same base fluid shows that in the case of the Ni_3-water-glycerol microfluid, a smaller value is achieved (Figure 14.10).

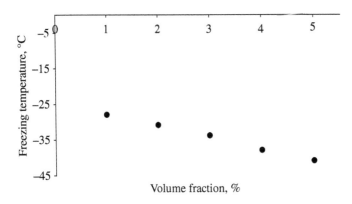

Figure 14.9 Concentration dependence of freezing point of Ni_3-water-glycerol.

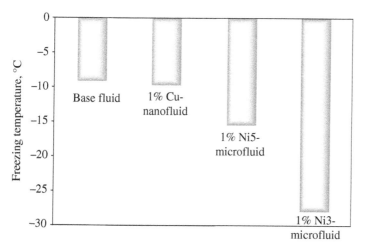

Figure 14.10 The freezing temperatures of the systems studied.

14.2.8 Concluding Remarks

1) The high thermal-conductivity enhancement achieved with Ni_3 MSC microparticles is due to higher stability of a microcolloidal system. The stability is provided by the formation of hydrogen bonds between the MSC particles (through their organic fragments) and water molecules and to the formation of microparticles assemblies.
2) The Ni_3 microfluid complex has a thixotropic rheology due to its colloidal structure and this structure has a significant impact on its thermophysical properties.
3) With a 5% volume fraction of Ni_3 MSC particles, the following properties of Ni_3 microfluid are observed: high thermal-conductivity enhancement (by 72%), low freezing point ($-41\,°C$), and low surface tension ($50\,mN\,m^{-1}$) as compared to the base fluid.
4) It is shown that thermal-conductivity enhancement of Ni_3 MSC microfluid is higher compared to other metal nanofluids.

14.3 Thermophysical Properties of Nano- and Microfluids with $Ni_5(\mu 5\text{-pppmda})_4Cl_2$ Metal String Complex Particles

Nanofluids are widely used in various industries for different applications [20, 41–46]. One such application is as a coolant in automobile radiators [20, 42, 44]. If the efficiency of the fluid could be increased then the amount of liquid needed could be reduced, which allows for smaller size and better positioning of the radiator in a vehicle [20, 42–44]. Additionally, coolant pumps could be shrunk and car engines operated at higher temperatures. These benefits could potentially be realized by using nanofluids as coolants.

The thermal conductivity, rheology, and stability of nanofluids are being broadly researched [41], but still raise many questions. There is still no agreed mechanism for their thermal conductivity enhancement. However, it is well known that the change in the thermophysical properties of nanofluids compared to microfluids is determined by their higher stability [47].

The theory of colloid stability and coagulation of ion-stabilized dispersed systems (DLVO theory) considers van der Waals–London molecular forces and the electrostatic force of repulsion arising from the overlap of similarly charged ionic atmospheres [48]. Deryaguin showed that under certain conditions on the potential curve of particle interaction, while maintaining the barrier of repulsion at a relatively far distance from the surface, a sufficiently deep secondary minimum occurs to allow loosely associated aggregates to form. This state of the system is a metastable equilibrium, the persistence of which is determined by the height of the barrier and the depth of the minimum. This association can allow the formation of periodic colloidal structures [35], which have a significant impact on the stability of nano and microfluids. Typically, stability is determined by the kinetics of changes in their density and rheological properties [49, 50].

Until now, the influence of monocrystalline nano- and microparticles on the thermophysical properties of fluids has not been considered. The role of the monocrystalline structure can be substantial, as increasing numbers of scattering centers that represent the boundaries between monocrystals constituting the polycrystal increase the residual thermal resistance and cause a reduction in the thermal conductivity of polycrystalline compared to monocrystalline metals [51–53].

In the following study, a detailed investigation of the stability and thermophysical properties of micro- and nanofluids containing the monocrystalline pentanickel (II) MSC microparticles and polycrystalline Cu or Al nanoparticles is presented.

14.3.1 Microparticles of the Metal String Complex $Ni_5(\mu_5\text{-pppmda})_4Cl_2$

Microparticles were synthesized for the first time by the reaction of anhydrous $NiCl_2$ with the N^2-(pyridin-2-yl)-N^6-(pyrimidin-2-yl)pyridine-2,6-diamine (H_2pppmda) ligand in an argon atmosphere employing naphthalene as the solvent and Bu^tOK as a base to deprotonate the amine group according to a previously reported method for other nickel string complexes, as described below [54–56].

Anhydrous $NiCl_2$ (0.78 g, 6.0 mmol), H_2pppmda (1.06 g, 4.0 mmol), and naphthalene (40 g) were placed in an Erlenmeyer flask. The mixture was heated to about 180–190 °C for 15 hours under argon and then a solution of potassium *tert*-butoxide (1.00 g, 8.91 mmol) in *n*-butyl alcohol (5 ml) was added dropwise. The mixture was refluxed for eight hours continuously. After the mixture had cooled, hexane was added to wash out naphthalene and then 200 ml of CH_2Cl_2 was used to extract the complex. A dark brown–purple complex was obtained after evaporation. The dark-brown block-shaped single crystals suitable for X-ray diffraction were obtained from diffusion of ether to a chloroform solution (0.78 g, 55% yield).

An X-ray crystal structure analysis revealed that, similar to other oligo-α-pyridylamino MSCs, the pentanickel linear metal chain in $Ni_5(\mu_5\text{-pppmda})_4Cl_2$ is helically wrapped by four *syn–syn–syn–syn* type pppmda^{2-} ligands. All of the nickel ions and two axial ligands are nearly collinear with the Ni–Ni–Ni angles in the range 178.13(4)–179.93(5) Å. The whole length of the Ni_5 chain is about 9.329 Å (Figure 14.11). The bulk density is about 1.668 g cm^{-3}.

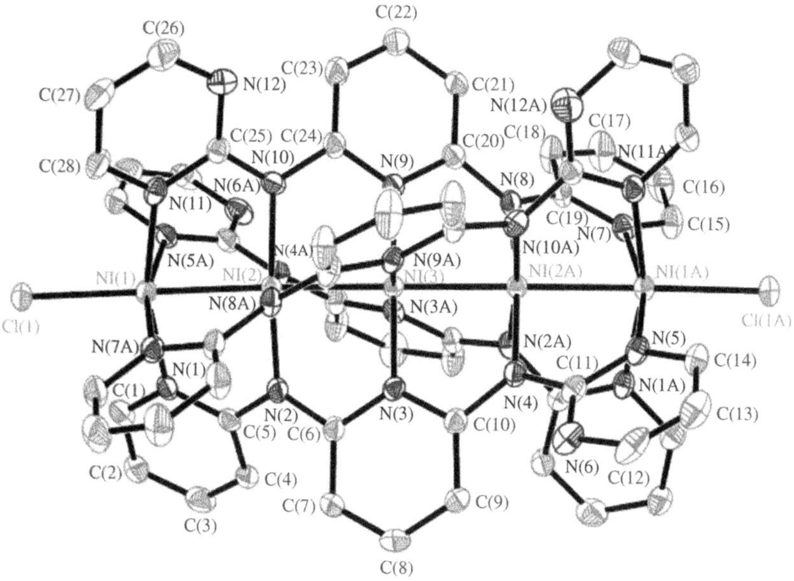

Figure 14.11 The crystal structure of $Ni_5(\mu_5\text{-pppmda})_4Cl_2$ (1). Thermal ellipsoids are at the 30% probability level and hydrogen atoms are omitted for clarity. Label A was generated through the symmetry operation $(-x, y, -z + 1/2)$.

The detailed crystal data, magnetic, and electrochemistry properties of this complex will be subject of another paper.

Microparticles of Ni were supplied by Sigma Aldrich and had an average size of 1 μm. The bulk density was about $3.5\,\mathrm{g\,cm^{-3}}$.

14.3.2 Micro- and Nanofluid Preparations

Surfactant solution: The sodium 4-alkyl-2yl-benzenesulfonate was used as a stabilizer for the copper nanoparticles and the polycrystalline nickel microparticles [57]. The necessary volume of this anionic surfactant for stabilizing the nanoparticles in a base liquid was estimated by the proposed formula, which assumes that the nanoparticles have a spherical shape:

$$\varphi' = 4\pi r_n^2 \delta n_n = 4\pi r_n^2 \delta \frac{\varphi}{V_n} = \frac{4\pi r_n^2 \delta \varphi}{\frac{4\pi}{3} r_n^3} = \frac{3\delta}{r_n}\varphi \qquad (14.3)$$

where φ is a volume fraction of the nanoparticles ($\varphi = n_n V_n$); n_n is the concentration of particles per unit volume; δ is the film thickness of surfactant; r_n is the radius of the nanoparticle; and V_n is the volume of the nanoparticle.

Base fluid: For carrying out experiments, a sufficient amount of a mixture of 70 wt% distilled water and 30 wt% glycerol was prepared as the base fluid. The density of the base fluid at 20 °C was about $1.07\,\mathrm{g\,cm^{-3}}$.

Nanofluids: *Cu-nanofluid*. Cu-nanoparticles (60 nm) were dispersed in the base fluid. *Al-nanofluid*. Al-nanoparticles (60 nm, 100 nm) were dispersed in the base fluid.

Microfluids: Ni5-microfluid. Ni5-microparticles were dispersed in the base fluid. Monocrystalline MSC (Ni$_5$(μ$_5$-pppmda)$_4$Cl$_2$) microparticles were obtained by a mechanical ball milling method, as described below. DLS data show that the size of the monocrystalline MSC (Ni$_5$(μ$_5$-pppmda)$_4$Cl$_2$) microparticles in the base fluid is around 1 μm and they are monodispersed in size, with a polydispersity of around 8%. *Ni-microfluid*. Polycrystalline Ni microparticles were dispersed in the base fluid solution. A Planetar PM 100 Ball Mill from the Retch Company was used. In addition to the well-proven mixing and size reduction processes, this mill also meets all of the technical requirements for colloidal grinding and provides the energy input necessary for mechanical alloying.

14.3.3 Fluid Stability

Particle size distribution of the nano- and microfluids were measured using a Horiba LB-550 Dynamic Light Scattering Particle Size Distribution Analyzer. The morphology and size studies of micro- and nanoparticles were performed using a JEOL JSM-7600F Scanning Electron Microscope.

SEM micrographs of the Cu-nanofluid (Figure 14.12a, b) clearly show the presence of non-homogenous aggregates of copper nanoparticles. However, the SEM micrographs of Ni$_5$ microfluid in Figure 14.12c, d, show a stable colloidal dispersion even after six hours. After sonication the Cu-nanofluid (at a concentration of 1 v. %) required additional treatments to stabilize it (Figure 14.13a). Sonication of Ni$_5$ microfluid led to additional milling of the particles (Figure 14.13b). For this reason, the monocrystalline metal string particles were well distributed in the base fluid and needed no additional treatments to stabilize them.

There are two reasons for the better stabilization of the Ni$_5$ complex microparticles in the base fluid:

- *Density*. The density of Ni$_5$ complex microparticles ($\rho_m = 1.668$ g cm^{-3}) is much smaller than the Cu nanoparticles ($\rho_{Cu} = 5$ g cm^{-3}), though greater than the base fluid ($\rho_{bf} = 1.07$ g cm^{-3}).
- *Association*. The SEM micrographs (Figure 14.12e) clearly show the formation of microparticle assemblies whose dimensions are determined by the equilibrium between gravity and buoyancy.

From Figure 14.12e it is seen that the average radius of the assembly (R_a) is 2.5 μm and the thickness of the associate surrounding liquid layer (H) is approximately 0.5 μm. According to reference [35], at large interparticle separations (compared to atoms), such as the 0.5 μm observed in our measurements, stable "secondary minimum" associations between colloidal particles can form with high degrees of order and a thixotropic colloidal structure.

It was necessary to evaluate the density of the microparticles (ρ_p) at which stabilization of the microfluid is achieved and assume their form to be spherical. According to Figure 14.12e, the average radius of the microparticles (r_p) can be taken equal to 0.5 μm and the average number of particles in the assembly (N) is approximately 10. The value of microparticle density at which microfluid stabilization is achieved can be determined from the condition of buoyancy, when the force of gravity acting on the assembly is balanced by the buoyancy:

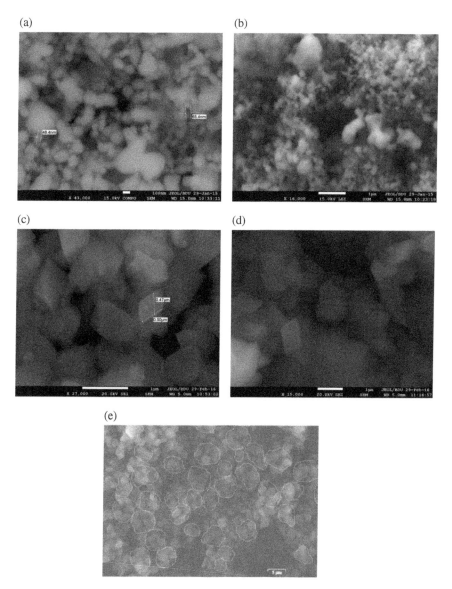

Figure 14.12 SEM images of Cu nanoparticles (1%) in glycerol: the magnification is (a) 43 000, (b) 13 000, and microparticles of monocrystalline metal string (Ni$_5$(μ_5-pppmda)$_4$Cl$_2$) (1%) in the base fluid: the magnification is (c) 27 000, (d) 15 000 after six hours of preparation, and (e) microparticle association.

$$N\frac{4}{3}\pi(r_p + H)^3 \rho_p g = \frac{4}{3}\pi R_a^3 \rho_{bf} g$$

where g is the acceleration due to gravity. The density of microparticles is calculated as follows:

$$\rho_p = \frac{\rho_{bf}}{N}\left(\frac{R_a}{r_p + H}\right)^3$$

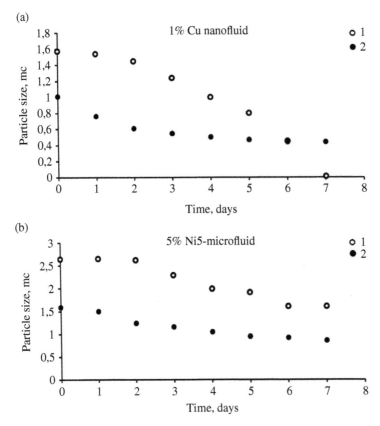

Figure 14.13 The effect of sonication on the stabilization of particle size for (a) copper (1%) and (b) monocrystalline metal string (Ni$_5$(µ$_5$-pppmda)$_4$Cl$_2$) complex (5%) in base fluid: (1) untreated dispersion, (2) after sonication.

Substituting in the formula the values of the parameters, we get $\rho_p = 1.672$ g cm^{-2}, which is practically the same as the density of Ni$_5$ complex particles.

14.3.4 Thermal Conductivity

A transient hot wire method (THWM) was used for thermal conductivity measurements of the investigated systems. This is based on the measurement of the temperature rise dependence and the temperature versus log time is measured as a current passes through a thin electrically insulated wire immersed in the test liquid [58].

A special setup (Figure 14.4) of a central element of which there is a platinum wire with length $L = 95$ mm and electrical resistance $R \approx 0.4$ Ω was used for conductivity measurements. Accepted values of timescales related to the flow and heat transfer for this setup and tested fluids were as follows: $t_{min} \geq 0.7$ s, $t_{max} \leq 240$ s. All measurements were taken at a constant heat output per unit length of wire equal to $I^2R/L \approx 4.21$ W m^{-1}, with the current $I = 1$ A at room temperature. A DC power supply PS-3005D was used as a stabilized DC source. The temperature versus time dependence was measured by a thermocouple placed

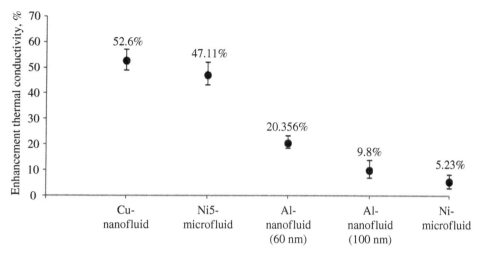

Figure 14.14 Thermal conductivity enhancement at the volume fraction of particles: $\varphi = 5\%$.

close to the middle of a platinum wire and the output was recorded by transmission of a signal from the thermocouple through the multimeter MT-1860 to the computer using an RS232 interface. Data transmitted to the computer were processed and the graphs of temperature versus log time $T(\ln t)$ for each sample were constructed. The thermal conductivity of the corresponding fluid was determined from the slope of the straight-line portion of the graph (s), compared with the reference fluid slope: $k_f = k_{et} S_{et}/S_f$. Distilled water at room temperature was taken as the reference fluid with $k_{et} = 0.6$ W (m K)$^{-1}$.

The thermal conductivity enhancement of nanofluids was studied up to the volume fraction of 5%, because further increasing the concentration of nanoparticles leads to instability and sedimentation. The results obtained are shown in Figure 14.14. It has been found that the thermal conductivity enhancements were 53% for the Cu-nanofluid and 47% for the Ni$_5$-microfluid in comparison with the base fluid. The results were relatively lower for the Al-nanofluids and Ni-microfluid: approximately 20% for one Al-nanofluid (60 nm), 10% for the other Al-nanofluid (100 nm), and only about 5% for the Ni-microfluid. It is necessary to note that the choice of the Ni microparticles instead of the Ni-nanoparticles was made for a comparison of the impact of the monocrystalline structure with the polycrystalline one having a similar dispersion.

Next, a series of experiments was carried out for the Cu-nanofluid and Ni$_5$-microfluid to study the impact of the volume fraction in solution on the thermal conductivity enhancement. The results presented in Figure 14.14 show an obviously linear dependence of these parameters.

The results can be explained within the framework of the proposed model for the mechanism of thermal conductivity enhancement by nanofluids [31]. According to this model, single nanoparticles (n), clusters (cl) formed by nanoparticles aggregation, and base fluid molecules contribute to the rate of heat transfer (Q) through the nanofluid according to Fourier's law:

$$Q = -\left(k_n A_n + k_n A_{cl} + k_f A_f\right)\frac{\partial T}{\partial r} = -k_f A_f \left(1 + \frac{k_n}{k_f}\frac{(A_n + A_{cl})}{A_f}\right)\frac{\partial T}{\partial r}$$

where k_n is the thermal conductivity coefficient of nanoparticles; k_f is the thermal conductivity coefficient of the base fluid molecules; A_n, A_{cl}, A_f are cross-sectional areas of single nanoparticles, clusters, and base fluid molecules adjacent to the cylindrical surface of the wire; and $\partial T/\partial r$ is the temperature gradient along the normal direction to the cylindrical surface.

By developing this model it was observed that the thermal conductivity enhancement of the nanofluid for small nanoparticle concentrations is in direct proportion to the volume fraction φ and is inversely proportional to the radius of nanoparticles r_n and aggregation ratio c (Figure 14.15):

$$\frac{k_{ef} - k_f}{k_f} = \frac{k_n}{k_f} \frac{\varphi r_f \left[1 + c\left(\frac{r_{cl}}{r_n}\right)^2\right]}{(1-\varphi) r_n \left[1 + c\left(\frac{r_{cl}}{r_n}\right)^3\right]} \tag{14.4}$$

where k_{ef} is the thermal conductivity coefficient of the nanofluid; k_f is the thermal conductivity coefficient of the base fluid; r_{cl} is the radius of a cluster formed by nanoparticle aggregation; and r_f is the radius of a base fluid molecule.

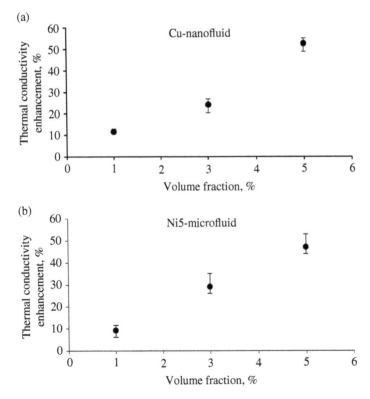

Figure 14.15 The dependence of thermal conductivity enhancement of the Cu-nanofluid (a) and the Ni$_5$-microfluid (b) on particle volume fractions.

It has been observed that the application of Ni$_5$ microparticles leads to almost the same degree of thermal conductivity enhancement for the microfluids as for the Cu-nanofluid. It was suggested that observed phenomena were caused by monocrystallinity of the nickel complex particles. In the case of polycrystalline Ni microparticles the thermal conductivity enhancement was only 5.23% and the system was unstable due to rapid sedimentation.

It is well known that heat is transferred in metals by conduction electrons (k_e) and phonons (k_g):

$$k = k_e + k_g$$

In pure metals the contribution of the last term is insignificant and the thermal conductivity of the electronic component can be represented as

$$k_e = W_e^{-1} = (W_0 + W_i)^{-1}$$

where W_e is the electronic thermal resistance, W_0 is the residual thermal resistance, due to crystal imperfections, and W_i is the own thermal resistance, caused by scattering of the conduction electrons by phonons in an electron–phonon interaction.

A greater number of scattering centers in the polycrystalline metal (i.e. grain boundaries between monocrystalline units) raise the contribution of residual thermal resistance to the total thermal resistance. The scattering of the conduction electrons in grain boundaries leads to a decrease in the thermal conductivity of the polycrystalline metal in comparison with the monocrystalline one [52, 59].

On the basis of the acquired results, it has been predicted that very high thermal conductivity enhancement (about 1 μm/100 nm ~ 10 times in accordance with Equation (14.4)) could be achieved by application of monocrystalline Ni$_5$ nanoparticles. However, the synthesis of monocrystalline MSC nanoparticles remains a challenge.

14.3.5 Rheology

The rheology of the investigated systems was studied on Brookfield Model BEL-PVSR-230 Rheometer, which is designed to test small complex samples by simulating process conditions in a bench-top environment and can also be used to measure the temperature dependence of fluid viscosity.

14.3.6 Surface Tension

The surface tension of the investigated systems was performed on a KRUSS DSA30 tensiometer using the pendant drop method.

Since the practical application of micro- and nanofluids is always associated with their flow, the rheological properties of Cu-nanofluid and Ni$_5$-microfluid were also studied. The results of rheology measurements are shown in Figures 14.16 and 14.17. It is obvious that the base fluid is Newtonian, but nanoparticle addition changes its rheological behavior to a pseudo-plastic. The rheology of the nanofluids studied is well described by the Ostwald-de Waele power law:

$$\tau = K\dot{\gamma}^n$$

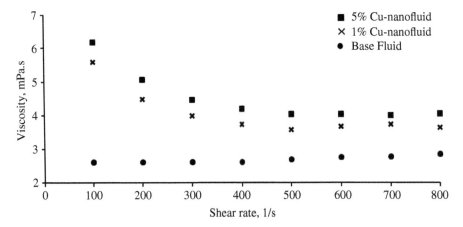

Figure 14.16 The dependence of viscosity on the shear rate for the Cu-nanofluid at 22 °C.

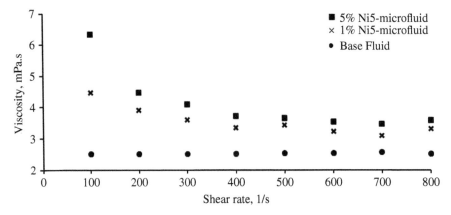

Figure 14.17 The dependence of viscosity on the shear rate for the Ni$_5$-microfluid at 25 °C.

where τ is shear stress, $\dot{\gamma}$ is shear rate, and K and n are constants that depend on the type of fluid.

The dependence of viscosity on the volume fraction for the Cu-nanofluid is shown in the Figure 14.18. As seen from this figure, there is a quadratic dependence of viscosity on the volume fraction [60].

The mechanism of viscosity enhancement for nanofluids in comparison with the base fluid is quite complex [61–64]. Some authors argue that the presence of nanoparticles in the liquid increases the average interaction radius between the molecules of the liquid and leads to "structuring" [63]. The nanofluid becomes more ordered than the base liquid and the degree of ordering increases with particle concentration. The characteristic linear scale of short-range molecular order near nanoparticles becomes approximately double that of the base fluid. An increase in the degree of ordering of the fluid increases its effective viscosity. A suggested explanation for this was that at low shear rates the nanoparticles

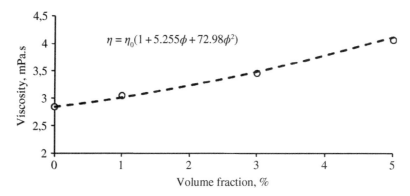

Figure 14.18 The dependence of the Cu-nanofluid viscosity on the volume fraction of copper nanoparticles.

in addition to translational motion are able to rotate, a fact that significantly contributes to the viscosity enhancement of nanofluids.

A number of studies found that the viscosity of nanofluids increases with particle size reduction [64, 65]. This happens due to the inversely proportional increase of surface area with a decrease in the nanoparticles radius (r) (i.e. spherical particles):

$$\frac{nS}{nV} \approx \frac{4\pi r^2}{\frac{4}{3}\pi r^3} = \frac{3}{r}$$

where n – is the number of nanoparticles per unit volume. Thus the internal friction to the fluid flow increases with increases in the specific surface area [65]. This was also confirmed by our experiments (Figure 14.19).

The behavior of the viscosity in the case of the associated Ni_5 complex (Figure 14.17) is determined by the thixotropic colloidal structure. In such systems the increase of shear rate destroys the colloidal structure, resulting in viscosity reduction [40, 66].

As monocrystalline metal string particles were well distributed in the base fluid, there is no need for additional treatments to stabilize them. Some further work was aimed to stabilize the Cu-nanoparticles in the base fluid.

Surfactant addition reduces the surface tension of nanofluids by adsorption of surfactant molecules at the liquid-nanoparticle interface which prevents their association and leads to nanofluid stability. The best result for Cu-nanoparticles stability in the base fluid was achieved by using 0.1 wt% of surfactant (formula (14.3)). The surfactant reduces the surface tension of the base liquid by adsorption at the interface which is a desirable effect for enhanced oil recovery (Figure 14.20).

According to the "hole" theory of viscous liquid flow [67], the square of the viscosity is inversely proportional to the surface tension at a given temperature, which agrees with the obtained experimental results (Figures 14.16, 14.17, and 14.19):

$$\eta = 0.915 \frac{RT}{V} \sqrt{\frac{m}{\sigma}} e^{\Delta E/RT}$$

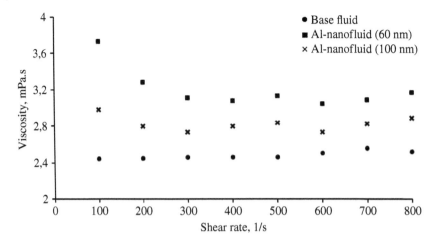

Figure 14.19 The dependence of viscosity on the shear rate for the Al-nanofluid at a temperature of 25 °C.

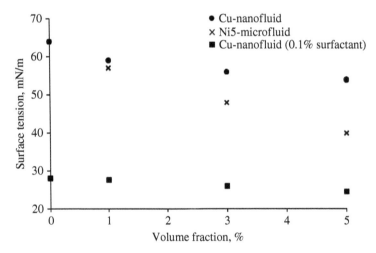

Figure 14.20 The dependence of the Cu-nanofluid surface tension on the volume fraction of particles with/without surfactant and for the Ni_5-microfluid.

where R is the universal gas constant; m is the mass of the fluid molecule; σ is the surface tension; V is the molar volume of the liquid; ΔE is the activation energy; and T is the absolute temperature.

Measurements made on the nano- and microfluids show that the addition of nano- and microparticles to the base fluid reduces microfluid surface tension (Figure 14.19). This is evident in the case of the Ni_5 microfluid. Addition of 5% of copper nanoparticles reduces the surface tension of the liquid by 15.6%, addition of the same quantity of Ni_5 microparticles in the same liquid leads to a 37.5% surface tension reduction. The strong decrease of surface tension also indicates that Ni$_5$ microparticles were better distributed throughout the volume and liquid surface [68].

14.3.7 Freezing Points

As noted above, nanofluids can potentially be successfully used as coolants in automotive radiators. Therefore, the freezing temperature was measured for the investigated systems (Figure 14.21).

Nanoparticle addition reduces the freezing temperature of the studied fluids. Apparently, the Ni$_5$ microparticles, being well distributed between the liquid molecules, prevent their convergence while freezing. It was apparent that the reason for this was the relative stability of the Ni$_5$ nanoparticles in comparison with other options, i.e. the impact of the Cu nanoparticles on the freezing point was not large due to the rapid agglomeration and further sedimentation of these particles (Figures 14.13a and 14.21).

The results obtained also confirm a stable colloidal structure of the Ni$_5$ microfluid. Uniformly distributed in base fluid, these microparticles and particle assemblies interfere with the convergence of the liquid molecules during cooling, which leads to a decrease in the freezing temperature (Figure 14.21) [49, 69].

14.3.8 Concluding Remarks

1) It is shown that Ni$_5$ microfluid has a higher stability than the other nanofluids tested. The high stability is based on the lower density of the microparticles and formation of microparticle assemblies. The assembly dimensions are determined by the equilibrium of gravity and buoyancy forces. The Ni$_5$ microfluid complex has a thixotropic rheology due to the microfluid colloidal structure.
2) The colloidal structure of the Ni$_5$ complex microfluid has a significant impact on its thermophysical properties.

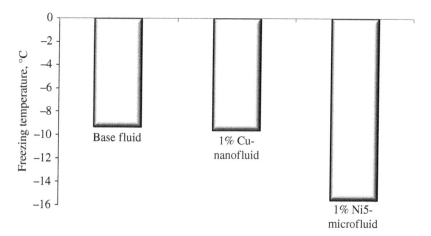

Figure 14.21 The freezing temperatures of the fluids studied.

3) It has been shown that thermal conductivity enhancement of microfluids is based on the monocrystalline structure of the MSC microparticles used. In the polycrystalline metal case the high concentration of scattering centers (i.e. grain boundaries between monocrystalline units) raises the contribution of residual thermal resistance to the total thermal resistance. Scattering of the conduction electrons in grain boundaries leads to decreased thermal conductivity of a polycrystalline metal in comparison with a monocrystalline one.
4) The developed theory predicts that very high thermal conductivity enhancement (up to 10 times) could be achieved if Ni_5-MSC microparticles are replaced by nanosized ones. Synthesis of Ni_5-MSC nanoparticles is the authors' future challenge.

Nomenclature

A_n, A_{cl}, A_f	cross-sectional areas of single nanoparticles, clusters and base fluid molecules
$a\ k/c\rho$	thermal diffusivity of fluid, m² s⁻¹
c	specific heat of fluid, J (kg K)⁻¹
C	aggregation ratio
G	gravitational acceleration, m s⁻²
H	thickness of the associate surrounding liquid layer, m
I	DC current, A
K	thermal conductivity coefficient of the homogeneous fluid, W (m K)⁻¹
k_{ef}	dispersion thermal conductivity coefficient, W (m K)⁻¹
k_f	base-fluid thermal conductivity coefficient, W (m K)⁻¹
k_n	particle thermal conductivity coefficient, W (m K)⁻¹
L	platinum wire length, m
N	average number of particles in the assembly
n_n	concentration of particles per unit volume
R	radial distance, m
r_{cl}	radius of a cluster formed by particle aggregation, m
r_f	radius of a base fluid molecule, m
r_n	particle radius, m
q	thermal flux per unit wire length, W m⁻¹
R	platinum wire resistance, Ω
R_a	average radius of the assembly, m
r_p	average radius of the microparticles, m
s	slope of the linear-dependence graph $T(\ln t)$
S_{et}	graph slope for reference fluid
S_f	graph slope for investigated fluid
S	wetting, N m⁻¹
T	temperature, °C or K
T_0	initial temperature, °C or K
T	time, s
V_n	volume of the nanoparticle

Nanocolloids for Petroleum Engineering: Fundamentals and Practices, First Edition.
Baghir A. Suleimanov, Elchin F. Veliyev, and Vladimir Vishnyakov.
© 2022 John Wiley & Sons Ltd. Published 2022 by John Wiley & Sons Ltd.

Greek Symbols

ΔT	temperature rise, °C or K
γ_l	surface tension of liquid, N m^{-1}
γ_s	surface tension of solid, N m^{-1}
γ_{sl}	interfacial tension between solid and liquid, N m^{-1}
$\gamma = 0.5772157$	Euler's constant
ρ	fluid density, kg m^{-3}
ρ_{bf}	density of base fluid, kg m^{-3}
ρ_{Cu}	density of copper nanoparticles, kg m^{-3}
ρ_m	density of complex microparticles, kg m^{-3}
ρ_p	particle density, kg m^{-3}
φ	particle volume fraction
δ	film thickness of surfactant

References

1 Reed, M.A. and Tour, J.M. (2000). Computing with molecules. *Scientific American* 282 (6): 86–93.
2 Wires, M. (2005). From design to perspectives: L. De Cola. *Topics in Current Chemistry* 257.
3 Krogmann, K. (1969). Planare komplexe mit metall-metall-bindungen. *Angewandte Chemie* 81 (1): 10–17.
4 Liu, I.P.C., Chen, C.H., and Peng, S.M. (2012). The road to molecular metal wires: the past and recent advances of metal string complexes. *Bulletin of Japan Society of Coordination Chemistry* 59: 3–10.
5 Berry, J.F. (2010). Metal–metal bonds in chains of three or more metal atoms: from homometallic to heterometallic chains. In: *Metal-Metal Bonding*, 1–28. Berlin, Heidelberg: Springer.
6 Kuo, C.K., Chang, J.C., Yeh, C.Y. et al. (2005). Synthesis, structures, magnetism and electrochemical properties of triruthenium–acetylide complexes. *Dalton Transactions* 22: 3696–3701.
7 Wang, W.Z., Wu, Y., Ismayilov, R.H. et al. (2014). A magnetic and conductive study on a stable defective extended cobalt atom chain. *Dalton Transactions* 43 (16): 6229–6235.
8 Yin, C., Huang, G.C., Kuo, C.K. et al. (2008). Extended metal-atom chains with an inert second row transition metal: [Ru5 (μ5-tpda) 4X2] (tpda2 = tripyridyldiamido dianion, X = Cl and NCS). *Journal of the American Chemical Society* 130 (31): 10090–10092.
9 Liu, I.P.C., Wang, W.Z., and Peng, S.M. (2009). New generation of metal string complexes: strengthening metal–metal interaction via naphthyridyl group modulated oligo-α-pyridylamido ligands. *Chemical Communications* 29: 4323–4331.
10 Suleimanov, B.A., Abbasov, H.F., Valiyev, F.F. et al. (2019). Thermal-conductivity enhancement of microfluids with Ni_3 (μ3-ppza) $4Cl_2$ metal string complex particles. *Journal of Heat Transfer* 141 (1).
11 Suleimanov, B.A., Ismayilov, R.H., Abbasov, H.F. et al. (2017). Thermophysical properties of nano-and microfluids with [Ni_5 (μ5-pppmda) $4Cl_2$] metal string complex particles. *Colloids and Surfaces A: Physicochemical and Engineering Aspects* 513: 41–50.
12 Maxwell, J.C. (1881). *A Treatise on Electricity and Magnetism*. Oxford: Clarendon Press.
13 Choi, S.U.S. (1995). Enhancing thermal conductivity of fluids with nanoparticles. In: *Developments and Applications of Non-Newtonian Flows*, vol. 231/MD–66 (ed. D.A. Siginer and H.P. Wang). New York: ASME.

Nanocolloids for Petroleum Engineering: Fundamentals and Practices, First Edition.
Baghir A. Suleimanov, Elchin F. Veliyev, and Vladimir Vishnyakov.
© 2022 John Wiley & Sons Ltd. Published 2022 by John Wiley & Sons Ltd.

14 Das, S.K., Choi, S.U.S., Wenhua, Y., and Pradeep, T. (2007). *Nanofluids: Science and Technology*. Hoboken, New Jersey: Wiley.

15 Choi, S.U.S. (2008). Nanofluids: A new field of scientific research and innovative applications. *Heat Transfer Engineering* 29: 429–431.

16 Solangi, K.H., Kazi, S.N., Luhur, M.R. et al. (2015). A comprehensive review of thermophysical properties and convective heat transfer to nanofluids. *Energy* 89: 1065–1086.

17 Jama, M., Singh, T., Gamaleldin, S.M. et al. (2016). Critical review on nanofluids: preparation, characterization, and applications. *Journal of Nanomaterials* 6717624.

18 Saidur, R., Leong, K.Y., and Mohammad, H.A. (2011). A review on applications and challenges of nanofluids. *Renewable and Sustainable Energy Reviews* 15: 1646–1668.

19 Colangelo, G., Favale, E., Milanese, M. et al. (2017). Cooling of electronic devices: Nanofluids contribution. *Applied Thermal Engineering* 127: 421–435.

20 Bhogare, R.A. and Kothawale, B.S. (2013). A review on applications and challenges of nanofluids as coolant in automobile radiator. *International Journal of Environmental Research and Public* 3: 1–11.

21 Keblinski, P., Phillpot, S.R., Choi, S., and Eastman, J.A. (2002). Mechanisms of heat flow in suspensions of nano-sized particles (nanofluids). *International Journal of Heat and Mass Transfer* 45: 855–863.

22 Yu, W. and Choi, S.U.S. (2003). The role of interfacial layers in the enhanced thermal conductivity of nanofluids: A renovated Maxwell model. *Journal of Nanoparticle Research* 5: 167–171.

23 Feng, Y., Yu, B., Xu, P., and Zou, M. (2007). The effective conductivity of nanofluids based on the nanolayer and the aggregation of nanoparticles. *Journal of Physics D: Applied Physics* 3164–3171.

24 Jang, S.P. and Choi, S.U.S. (2004). Role of Brownian motion in the enhanced thermal conductivity of nanofluids. *Applied Physics Letters* 84: 4316–4318.

25 Sundar, L.S., Farooky, M.H., Sarada, N., and Singh, M.K. (2013). Experimental thermal conductivity of ethylene glycol and water mixture based low volume concentration of Al_2O_3 and CuO nanofluids. *International Communications in Heat and Mass Transfer* 41: 41–46.

26 Colangelo, G., Favale, E., Milanese, M. et al. (2016). Experimental measurements of Al_2O_3 and CuO nanofluids interaction with microwaves. *Journal of Energy Engineering* 143 (2).

27 Colangelo, G., Favale, E., Paola, M. et al. (2016). Thermal conductivity, viscosity and stability of Al_2O_3-diathermic oil nanofluids for solar energy systems. *Energy* 95: 124–136.

28 Milanese, M., Iacobazzi, F., Colangelo, G., and de Risi, A. (2016). An investigation of layering phenomenon at the liquid-solid interface in Cu and CuO based nanofluids. *International Journal of Heat and Mass Transfer* 103: 564–571.

29 Iacobazzi, F., Milanese, M., Colangelo, G. et al. (2016). An explanation of the Al_2O_3 nanofluid thermal conductivity based on the phonon theory of liquid. *Energy* 116: 786–794.

30 Colangelo, G., Milanese, M., and de Risi, A. (2017). Numerical simulation of thermal efficiency of an innovative Al_2O_3 nanofluid solar thermal collector: influence of nanoparticles concentration. *Thermal Science* 21 (6): 2769–2779.

31 Suleimanov, B.A. and Abbasov, H.F. (2016). Effect of copper nanoparticle aggregation on the thermal conductivity of nanofluids. *Russian Journal of Physical Chemistry A* 90: 420–428.

32 Tsao, T.-B., Lee, G.-H., Yeh, C.-Y., and Peng, S.-M. (2003). Supramolecular assembly of linear trinickel complexes incorporating metalloporphyrins: a novel one-dimensional polymerand oligomer. *Dalton Transactions* 8: 1465–1471.

33 Clerac, R., Cotton, F.A., Dunbar, K.R. et al. (1999). Further study of the linear trinickel(ii) complex of dipyridylamide. *Inorganic Chemistry* 38: 2655–2657.

34 Ismayilov, R.H., Wang, W.-Z., Lee, G.-H. et al. (2007). New versatile ligand family, pyrazine-modulated oligo-α-pyridylamino ligands, from coordination polymer to extended metal atom chains. *Dalton Transactions* 27: 2898–2907.

35 Efremov, I.F. and Usyarov, O.G. (1976). The long-range interaction between colloid and other particles and the formation of periodic colloid structures. *Russian Chemical Reviews* 45: 435–453.

36 Nagasaka, Y. and Nagashima, A. (1981). Absolute measurement of the thermal conductivity of electrically conducting liquids by the transient hot-wire method. *Journal of Physics E: Scientific Instruments* 14: 1435–1440.

37 Hong, S.W., Kang, Y.T., Kleinstreuer, C., and Koo, J. (2011). Impact analysis of natural convection on thermal conductivity measurements of nanofluids using the transient hot-wire method. *International Journal of Heat and Mass Transfer* 54: 3448–3456.

38 Chen, I.W.P., Fu, M.D., Tseng, W.H. et al. (2006). Conductance and stochastic switching of ligand-supported linear chains of metal atoms. *Angewandte Chemie International Edition* 45: 5814–5818.

39 Zafarani-Moattar, M.T. and Majdan-Cegincara, R. (2013). Investigation on stability and rheological properties of nanofluid of ZnO nanoparticles dispersed in poly(ethylene glycol). *Fluid Phase Equilibria* 354: 102–108.

40 Tseng, W.J. and Lin, K.-C. (2003). Rheology and colloidal structure of aqueous TiO_2 nanoparticle suspensions. *Materials Science and Engineering A* 355: 186–192.

41 Das, S.K., Choi, C.U.S., and Patel, H.E. (2006). Heat transfer in nanofluids – a review. *Heat Transfer Engineering* 27: 3–19.

42 Bupesh Raja, V.K., Unnikrishnan, R., and Purushothaman, R. (2015). Application of nanofluids as coolant in automobile radiator – An overview. *Applied Mechanics and Materials* 766–767: 337–342.

43 Naraki, M., Peyghambarzadeh, S.M., Hashemabadi, S.H., and Vermahmoudi, Y. (2013). Parametric study of overall heat transfer coefficient of CuO/water nanofluids in acar radiator. *national Journal of Thermal Sciences* 66: 82–90.

44 Leong, K.Y., Saidur, R., Kazi, S.N., and Mamun, A.H. (2010). Performance investigation of a car radiator operator with nanofluid-based coolants (nanofluid as a coolant in a radiator). *Applied Thermal Engineering* 30: 2685–2692S.

45 Eastman, J.A., Choi, S.U.S., Yu, W., and Thompson, L.J. (2001). Anomalously increased effective thermal conductivities of ethylene glycol-based nanofluids containing copper nanoparticles. *Applied Physics Letters* 78: 718.

46 Sorbie K.S., Wat R.M.S., and Rowe T. (1987). Oil displacement experiments in heterogeneous cores. *SPE 62nd Annual Fall Conference*.

47 Li, Y., Zhou, J., Tung, S. et al. (2009). A review on development of nanofluid preparation and characterization. *Powder Technology* 196: 89–101.

48 Deryguin, B.V. (1986). *Theory of Stability of Colloids and Thin Films*. Moscow: Nauka.

49 Gao, J., Ndong, R., Shiflett, M.B., and Wagner, N.J. (2015). Creating nanoparticle stability in ionic liquid [C4mim][BF4] by inducing solvation layering. *ACS Nano* 9: 3243–3253.

50 Rao, Y. (2010). Nanofluids: stability, phase diagram, rheology and applications. *Particuology* 8: 549–555.

51 Wang, Z. (2012). Thermal conductivity of polycrystalline semiconductors and ceramics. University of California, http://escholarship.org/uc/item/06r8376h#page-1.

52 McLeod, A.D., Haggerty, J.S., and Sadoway, D.R. (1984). Electrical resistivities of monocrystalline and polycrystalline TiB_2. *Journal of the American Ceramic Society* 67: 705–708.

53 Greenstein, A.M., Graham, S., Hudiono, Y.C., and Nair, S. (2006). Thermal properties and lattice dynamics of polycrystalline MFI zeolite films. *Nanoscale and Microscale Thermophysical Engineering* 10: 321–331.

54 Ismayilov, R.H., Wang, W.Z., Lee, G.H., and Peng, S.M. (2015). Study on a new polydentatepyridyl amine and its complexes: Synthesis, supramolecular structure and properties. *SOCAR Proceedings* 1: 74–82.

55 Shieh, S.-J., Chou, C.-C., Lee, G.-H. et al. (1997). Line arpentanuclear complexes containing a chain of metal atoms: [Co(μ5-tpda)4(NCS)2] and [Ni(μ5-tpda)4Cl₂]. *Angewandte Chemie (International Ed. in English)* 36: 56–59.

56 Yin, C.-X., Huo, F.-J., Wang, W.-Z. et al. (2009). Synthesis, structure, magnetism and electrochemical properties of linear pentanuclear complex: Ni5(μ-dmpzda)4(NCS)2. *Chinese Journal of Chemistry* 27: 1295–1299.

57 Suleimanov, B.A., Ismailov, F.S., and Veliyev, E.F. (2011). Nanofluid for enhanced oil recovery. *Journal of Petroleum Science and Engineering* 78: 431–437.

58 Hong, S.W., Kang, Y.T., Kleinstreuer, C., and Koo, J. (2011). Impact analysis of naturalconvection on thermal conductivity measurements of nanofluids using thetransient hot-wire method. *International Journal of Heat and Mass Transfer* 54: 3448–3456.

59 Chae, D.-H., Berry, J.F., Jung, S. et al. (2006). Vibrational excitations in single trimetal-molecule transistors. *Nano Letters* 6: 165–168.

60 Mishra, P.C., Mukherjee, S., Nayak, S.K., and Panda, A.A. (2014). Brief review on viscosity of nanofluids. *International Nano Letters* 4: 109–120.

61 Barlaka, S., Sara, O.N., Karaipekli, A., and Yapici, S. (2016). Thermal conductivity and viscosity of nanofluids having nanoencapsulated phase change material. *Nanoscale and Microscale Thermophysical Engineering*.

62 Rudyak, V.Y. and Krasnolutskii, S.L. (2014). Dependence of the viscosity of nanofluids on nanoparticle size and material. *Physics Letters A* 378: 1845–1849.

63 Venerus, D.C., Buongiorno, J., Christianson, R. et al. (2010). Viscosity measurements on colloidal dispersions (nanofluids) for heat transfer applications. *Applied Rheology* 20: 44582.

64 Lu, W. and Fan, Q. (2008). Study for the particle's scale effect on some thermophysical properties of nanofluids by a simplified molecular dynamics method. *Engineering Analysis with Boundary Elements* 32: 282–289.

65 Anoop, K.B., Sundararajan, T., and Das, S.K. (2009). Effect of particle size on the convective heat transfer in nanofluid in the developing region. *International Journal of Heat and Mass Transfer* 52: 2189–2195.

66 Zafarani-Moattar, M.T. and Majdan-Cegincara, R. (2013). Investigation on stability andrheological properties of nanofluid of ZnO nanoparticles dispersed inpoly(ethylene glycol). *Fluid Phase Equilibria* 354: 102–108.

67 Fürth, R. (1941). On the theory of the liquid state. *Mathematical Proceedings of the Cambridge Philosophical Society* 37: 281–290.

68 Ravera, F., Santini, E., Loglio, G. et al. (2006). Effect of nanoparticles on interfacial properties of liquid/liquid and liquid/air surface layers. *The Journal of Physical Chemistry B* 110: 19543–19551.

69 Löwen, H. (1994). Melting, freezing and colloidal suspensions. *Physics Reports* 237: 249–324.

Appendix A

Determination of Dispersed-Phase Particle Interaction Influence on the Rheological Behavior

Example Let us consider the motion of a dispersion in a tube with radius R. Let N_1 and N_2 be the concentrations of elementary inclusions and associates respectively. The names "elementary inclusions" and "associates" are conventional; the former is understood to be the particles that are not broken under any experimental conditions, while the latter denotes larger particles formed as a result of aggregation of elementary inclusions. Then, on the basis of the Lotka–Volterra and Ferhülster models, neglecting, in the first approximation, the spatial inhomogeneity because of the relative smallness of the concentration of particles, the system of kinetic equations that describe the evolution of the dispersed phase may be presented in the following form:

$$\frac{dN_1}{dt} = (a_2\gamma - \beta_1)N_2 - a_1\gamma N(1 - \alpha_1 N_1) - (\beta_1 + \beta_2)N_1 + \beta_1 N$$

$$\frac{dN_2}{dt} = -a_2 N_2 \gamma(1 - \alpha_2 B N_2) + \beta_2 N_1, \quad N_1(0) + N_2(0) \leq N = \text{constant}$$
(A.1)

For an axisymmetrical case, the equation of liquid motion in a cylindrical tube has the form

$$\rho\frac{\partial u}{\partial t} = \eta\left[\frac{\partial^2 u}{\partial r^2} + \frac{1}{r}\frac{\partial u}{\partial r}\right] + \frac{\Delta P}{l}$$
(A.2)

In the first approximation, to estimate the influence of the dynamics of motion of the liquid itself on the system rheology we will assume that the rate of shear over the cross-section is $\gamma = -\partial u/\partial r = \text{constant}$. Then we have

$$u = \gamma(R - r)$$
(A.3)

From Equation (A.2), subject to Equation (A.3), with the assumption adopted, we obtain

$$\rho\frac{d\gamma}{dt}(R - r) = -\frac{\eta}{r}\gamma + \frac{\Delta P}{l}$$
(A.4)

Having multiplied each term of Equation (A.4) by $2\pi r\, dr$ and divided by πR^2, we average both of its sides over the cross-section of the tube:

$$\frac{d\gamma}{dt} = -\frac{6\eta}{\rho R^2}\gamma + 3\frac{\Delta P}{\rho l R}$$
(A.5)

Nanocolloids for Petroleum Engineering: Fundamentals and Practices, First Edition.
Baghir A. Suleimanov, Elchin F. Veliyev, and Vladimir Vishnyakov.
© 2022 John Wiley & Sons Ltd. Published 2022 by John Wiley & Sons Ltd.

The viscosity of the system η changes with time and can be determined from the formula (A.1) as

$$\eta = \eta_0 + \eta_1 \frac{N_2}{N} \tag{A.6}$$

From Equation (A.5), subject to Equation (A.6), we find that

$$\frac{d\gamma}{dt} = -\frac{6\gamma}{\rho R^2}\left(\eta_0 + \eta_1 \frac{N_2}{N}\right) + 3\frac{\Delta P}{\rho l R} \tag{A.7}$$

From the system of Equations (A.1) and (A.6), we may determine the dynamics of the change in the rheological parameters of the disperse system. The stationary values of the parameters N_1, N_2, and γ can be found from Equations (A.1) and (A.7), which yield

$$q_2 = \overline{N}_2 = \frac{\beta_1 + \frac{\beta_1}{\beta_2}a_2\gamma_0 + \frac{a_1 a_2}{\beta_2}\gamma_0^2 + \sqrt{\left(\beta_1 + \frac{\beta_1}{\beta_2}a_2\gamma_0 + \frac{a_1 a_2}{\beta_2}\gamma_0^2\right) - 4\beta_1 Na_0}}{2a_0} \tag{A.8}$$

$$a_0 = \frac{a_1 a_2 \gamma_0^2}{\beta_2}a_2 + \frac{a_1 a_2^2 a_1 \gamma_0^3}{\beta_2^2}a_2 + \frac{\beta_1 + \beta_2}{\beta_2}a_2\gamma_0 a_2, \quad q_1 = \overline{N}_1 = \frac{a_2 \gamma_0 q_2(1 - a_2 q_2)}{\beta_2} \tag{A.9}$$

$$\gamma_0 = \frac{\Delta P}{l} \frac{P}{2\left(\eta_0 + \eta_1 \frac{q_2}{N}\right)} \tag{A.10}$$

If α_1 and α_2 are much less than a_1 and a_2, then, from Equation (A.8), we obtain

$$q_1 = \frac{\frac{a_2 \gamma_0}{\beta_2}N}{1 + \frac{a_2}{\beta_2}\gamma_0 + \frac{a_1 a_2}{\beta_1 \beta_2}\gamma_0^2}, \quad q_2 = \frac{N}{1 + \frac{a_2}{\beta_2}\gamma_0 + \frac{a_1 a_2}{\beta_1 \beta_2}\gamma_0^2} \tag{A.11}$$

From Equation (A.10), subject to Equation (A.11), we have

$$\gamma_0^3 - \left(\frac{\Delta PR}{2\eta_0 l} - \frac{\beta_1}{a_1}\right)\gamma_0^2 - \left[\frac{\Delta PR}{2\eta_0 l}\frac{\beta_1}{a_1} - \left(1 + \frac{\eta_1}{\eta_0}\right)\frac{\beta_1 \beta_2}{a_1 a_2}\right]\gamma_0 - \frac{\Delta PR}{2\eta_0 l}\frac{\beta_1 \beta_2}{a_1 a_2} = 0 \tag{A.12}$$

Equation (A.12) yields

$$\gamma_0 = \sqrt[3]{-\frac{n}{2} + \sqrt{\left(\frac{n}{2}\right)^2 + \left(\frac{m}{3}\right)^3}} + \sqrt[3]{-\frac{n}{2} - \sqrt{\left(\frac{n}{2}\right)^2 + \left(\frac{m}{3}\right)^3}} + \frac{1}{3}\left(\frac{\Delta PR}{2\eta_0 l} - \frac{\beta_1}{a_1}\right) \tag{A.13}$$

where

$$m = -\frac{1}{3}\left(\frac{\Delta PR}{2\eta_0 l} - \frac{\beta_1}{a_1}\right)^2 - \frac{\Delta PR}{2\eta_0 l}\frac{\beta_1}{a_1} + \left(1 + \frac{\eta_1}{\eta_0}\right)\frac{\beta_1 \beta_2}{a_1 a_2}$$

and

$$n = -\frac{2}{27}\left(\frac{\Delta PR}{2\eta_0 l} - \frac{\beta_1}{a_1}\right)^2 - \frac{1}{3}\left(\frac{\Delta PR}{2\eta_0 l} - \frac{\beta_1}{a_1}\right)\left[\frac{\Delta PR}{2\eta_0 l}\frac{\beta_1}{a_1} - \left(1 + \frac{\eta_1}{\eta_0}\right)\frac{\beta_1 \beta_2}{a_1 a_2}\right] - \frac{\Delta PR}{2\eta_0 l}\frac{\beta_1 \beta_2}{a_1 a_2}$$

Appendix A Determination of Dispersed-Phase Particle Interaction Influence on the Rheological Behavior | 261

The values of q_1 and q_2 are determined from Equation (A.11), subject to Equation (A.13). In the first approximation, we assume that there are slight fluctuations of v_1, v_2, and ε around the stationary position of q_1, q_2, and γ_0:

$$N_1 = q_1(1+v_1), \quad N_2 = q_2(1+v_2), \quad \gamma = \gamma_0 + \varepsilon \tag{A.14}$$

Then, from Equations (A.1) and (A.7), with Equation (A.14) taken into account, we obtain an equation of a perturbed motion of the system from the first approximation:

$$\frac{dv_1}{dt} = -b_1 v_1 + \frac{q_2}{q_1}(a_2\gamma_0 - \beta_1)v_2 + \left(a_1 q_1 + \frac{a_2 q_2}{a_1 q_1} - 1\right) a_1 \varepsilon$$

$$\frac{dv_2}{dt} = \frac{q_1}{q_2} v_1 + (2a_2 q_2 - 1)a_2\gamma_0 v_2 + (a_2 q_2 - 1)a_3 \varepsilon$$

$$\frac{d\varepsilon}{dt} = -\frac{6}{\rho R^2}\gamma_0 \eta_1 \frac{q_2}{N} v_2 - \frac{6}{\rho R^2}\left(\eta_0 + \frac{\eta_1 q_2}{N}\right)\varepsilon \tag{A.15}$$

The threshold values of the parameters at which the loss of stability of the nonperturbed motion (Equations (A.1) and (A.2)) of the system occurs can be determined from Equation (A.15). For the equation of perturbed motion, (A.15), the characteristic equation has the form

$$\begin{vmatrix} -b_2 - \lambda & \frac{q_2}{q_1}(a_1\gamma_0 - \beta_1) & \left(a_1 q_1 + \frac{a_1 q_2}{a_1 q_2} - 1\right) a_1 \\ \frac{q_1}{q_2}\beta_2 & (2a_2 q_2 - 1)a_2\gamma_0 - \lambda & (a_2 q_2 - 1)\, a_3 \\ 0 & -\frac{6}{\rho R^2}\gamma_0 \eta_1 \frac{q_2}{N} & -\frac{6}{\rho R^2}\left(\eta_0 + \frac{\eta_1 q_2}{N}\right) - \lambda \end{vmatrix} = 0 \tag{A.16}$$

from which

$$\lambda^3 + c_1 \lambda^2 + c_2 \lambda + c_3 = 0 \tag{A.17}$$

where

$$c_1 = a_2\gamma_0(1 - 2a_2 q_2) + \frac{6}{\rho R^2}\left(\eta_0 + \frac{\eta_1 q_2}{N}\right) + b_1$$

$$c_2 = (1 - 2a_2 q_2)a_2\gamma_0 \frac{6}{\rho R^2}\left(\eta_0 + \frac{\eta_1 q_2}{N}\right)$$

$$+ b_1(1 - 2a_2 q_2)a_2\gamma_0 + b_1 \frac{6}{\rho R^2}\left(\eta_0 + \frac{\eta_1 q_2}{N}\right) + (1 - a q_2)a_3 \frac{6}{\rho R^2}\gamma_0 \eta_1 \frac{q_2}{N} + (a_2\gamma_0 - \beta_1)\beta_2 \frac{6}{\rho R^2}$$

$$c_3 = b_1(1 - 2a_2 q_2)a_2\gamma_0 \frac{6}{\rho R^2}\left(\eta_0 + \frac{\eta_1 q_2}{N}\right) + b_1(1 - a q_2)a_3 \frac{6}{\rho R^2}\gamma_0 \eta_1 \frac{q_2}{N}$$

$$+ (\beta_1 - a_2\gamma_0)\beta_2 \frac{6}{\rho R^2}\left(\eta_0 + \frac{\eta_1 q_2}{N}\right) + \left(a_1 q_1 + \frac{a_2 q_2}{a_1 q_1} - 1\right)\frac{a_1 q_1}{a_2 q_2}$$

$$b_1 = a_1\gamma_0(1 - 2a_1 q_1) + \beta_1 + \beta_2$$

The Hurwitz matrix for Equation (A.17) has the form

$$A = \begin{pmatrix} c_1 & 1 & 0 \\ c_3 & c_2 & c_1 \\ 0 & 0 & c_3 \end{pmatrix} \tag{A.18}$$

Appendix A Determination of Dispersed-Phase Particle Interaction Influence on the Rheological Behavior

According to the Lyapunov theorem, for asymptotic stability of nonperturbed equations of motion (Equations (A.1) and (A.2)) of the system, it is necessary and sufficient that Equation (A.17) could have negative real parts. Then, according to the Hurwitz criteria, all major diagonal minors of matrix (A.18) must be positive, i.e.

$$c_1 > 0, \quad c_1 c_2 c_3 > 0, \quad c_3(c_1 c_2 - c_3) > 0 \tag{A.19}$$

Moreover, according to the Lyapunov theorem, if at least one of the minors in Equation (A.18) is negative, then the nonperturbed motion of the system is unstable at any nonlinear terms on the right-hand sides of Equations (A.1) and (A.2). Thus, the resulting conditions in Equation (A.19) make it possible to determine the threshold values of the parameters at which an unstable state of the system sets in. Since the shear stress τ in the system can be determined from the formula

$$\tau = \eta \gamma \tag{A.20}$$

From Equation (A.20), subject to Equations (A.6), (A.11), and (A.14), for the limiting case $t \to \infty$ we have

$$\tau = \gamma_0 \left(\eta_0 + \frac{\eta_1}{1 + \frac{a_2}{\beta_2}\gamma_0 + \frac{a_1 a_2}{\beta_1 \beta_2}\gamma_0^2} \right) \tag{A.21}$$

Numerical calculation of the value of τ from Equation (A.20) has been performed at the following values of the parameters (that were determined from the experimental curves of the flow with the aid of the least-squares method): $\eta_0 = 4$ mPa s; $\eta_1 = 20$ mPa s; $a_1 = a_2 = 10^{-8}$; and $\beta_1 = \beta_2 = 10^{-6}$ s^{-1}. The results of calculation together with the results of experimental investigations are presented in Figure 2.10. As seen from the figure, the results of the theoretical investigations agree well with experimental data. At the above-indicated values of the parameters and a value of the rate of shear γ above 600 s^{-1}, the stability of the system is perturbed. Next, as shown in references [1–5], in a system similar to that under study, at values of the parameters above the threshold, chaos sets in because of the two-, three-, or fivefold increases in the period of fluctuations, which agrees with the results of the investigations carried out.

The results obtained can be used for hydrodynamic calculations in oil and gas production, as well as to describe filtration of non-Newtonian systems in porous media.

Appendix B

Determination of Inflection Points

Consider the logistics of order 2, defined by the equation

$$y(t) = \frac{K}{1 + b_1 e^{-a_1 t} + b_2 e^{-a_2 t}}, 0 < a_1 < a_2 \quad (B.1)$$

where K is the limit which tends the exponent at $t \to \infty$; a_i, b_i ($i = 1, 2$) are parameters determined according to statistical data (y_i, t_i), ($i = 1, ..., n$).

From the condition of a process increasing with time. it follows that the number of inflection points for function (B.1) can only be odd. The inflection points are found from the equation $y''(t) = 0$, which implies the transcendental equation

$$(a_1 b_1)^2 e^{-2a_1 t} + b_1 b_2 (4a_1 a_2 - a_1^2 - a_2^2) e^{-(a_1 + a_2)t} + (a_2 b_2)^2 e^{-2a_2 t} \\ - a_1^2 b_1 e^{-a_1 t} - a_2^2 b_2 e^{-a_2 t} = 0 \quad (B.2)$$

With sufficient accuracy for practical purposes, we can assume that a_1 and a_2 are rational numbers. Let $a_1 = m_1/n_1$, $a_2 = m_2/n_2$, where m_i, n_i ($i = 1, 2$) are positive integers. We denote N as the common denominator of partial fractions m_1/n_1 and m_2/n_2. Let $N_i = a_i N$ ($i = 1, 2$), $\tau = t/N$. Then Equation (B.2), after the replacement $x = e^{-\tau}$, reduces to the algebraic equation of $2N_2 - N_1$ degrees (because a higher $a_2 > a_1$ is accepted):

$$(N_2 b_2)^2 x^{2N_2 - N_1} + b_1 b_2 (4N_1 N_2 - N_1^2 - N_2^2) x^{N_2} + (N_1 b_1)^2 x^{N_1} - b_2 N_2^2 x^{N_2 - N_1} - b_1 N_1^2 = 0 \quad (B.3)$$

We assume that $b_1 > b_2 > 0$. We are interested only in the positive equation roots of Equation (B.3), since $e^{-\tau} > 0$. As shown in reference [6], Equation (B.3) has only one positive root if $N_1 < N_2 \leq 2N_1$ and one or three positive roots if $N_2 > 2N_1$. To find them we will proceed from the difference quotient

$$\frac{1}{y(t+2)} - \frac{e^{-a_1} + e^{-a_2}}{y(t+1)} + \frac{e^{-a_1 - a_2}}{y(t)} = \frac{1 - (e^{-a_1} + e^{-a_2}) + e^{-a_1 - a_2}}{K} \quad (B.4)$$

The validity of the quotient is easily seen by substituting $y(t)$ in it by formula (B.1). Let us denote

$$v_1 = e^{-a_1} + e^{-a_2}, v_2 = e^{-a_1 - a_2}, v_3 = (1 - v_1 + v_2)/K,$$

264 | Appendix B Determination of Inflection Points

$$X^{(1)} = 1/y(t+1), X^{(2)} = 1/y(t), Y = 1/y(t+2) - K, \tag{B.5}$$

A chart of the true function is shown in Figure 12.1. For convenience, the scale of t was increased by 100 times. From Figure 12.1, we find that $K = 48$.

Minimizing the expression

$$\sum_{t=1}^{n-2} \left[\left(\frac{1}{y(t+2)} - K \right) + \frac{v_1}{y(t+1)} \left(\frac{1}{K} - 1 \right) + \frac{v_2}{y(t_i)} \left(\frac{1}{K} + 1 \right) \right]^2. \tag{B.6}$$

Using the method of least squares (MLS), we find the values of the coefficients for regression:

$$Y = c_1 X^{(1)} + c_2 X^{(2)}, \tag{B.7}$$

where

$$c_1 = \left(1 - \frac{1}{K}\right) v_1, \quad c_2 = \left(1 + \frac{1}{K}\right) v_2. \tag{B.8}$$

MLS-estimations of the regression coefficients c_1, c_2:

$$\hat{c}_1 = 0.91868, \hat{c}_2 = -0.081708 \tag{B.9}$$

From Equations (B.8) and (B.9) we obtain

$$v_1 = \frac{\hat{c}_1}{\left(1 - \frac{1}{K}\right)}, v_2 = \frac{\hat{c}_2}{\left(1 + \frac{1}{K}\right)}. \tag{B.10}$$

and parameters a_1 and a_2 are calculated by the formulas

$$a_1 = \ln \frac{2}{v_1 + \sqrt{v_1^2 - 4v_2}}, a_2 = \ln \frac{2}{v_1 - \sqrt{v_1^2 - 4v_2}}. \tag{B.11}$$

From Equation (B.11) subject to Equations (B.9) and (B.10), we find $a_1 = 0.1704$ and $a_2 = 2.3548$, from whence it follows that $a_1 = N_1/N$, $a_2 = N_2/N$, $N_1 = 17$, $N_2 = 235$, and $N = 100$.

The remaining parameters b_1, b_2 are found once again by the least-squares method from the condition of the least sum

$$\sum_{i=1}^{n} \left[\frac{K}{y(t_i)} - 1 - b_1 e^{-a_1 t_i} - b_2 e^{-a_2 t_i} \right]^2. \tag{B.12}$$

This procedure allows b_1 and b_2 to be defined:

$$b_1 = \frac{\left[\sum_{i=1}^{n} e^{-a_1 t_i} \left(\frac{K}{y(t_i)} - 1 \right) \sum_{i=1}^{n} e^{-2a_2 t_i} - \sum_{i=1}^{n} e^{-a_2 t_i} \left(\frac{K}{y(t_i)} - 1 \right) \sum_{i=1}^{n} e^{-(a_1 + a_2) t_i} \right]}{\Delta_1} \tag{B.13}$$

$$b_2 = \frac{\left[\sum_{i=1}^{n} e^{-a_2 t_i} \left(\frac{K}{y(t_i)} - 1 \right) \sum_{i=1}^{n} e^{-2a_1 t_i} - \sum_{i=1}^{n} e^{-a_1 t_i} \left(\frac{K}{y(t_i)} - 1 \right) \sum_{i=1}^{n} e^{-(a_1 + a_2) t_i} \right]}{\Delta_1}$$

From Equations (B.13), we obtain

$$\hat{b}_1 = 589.78408, \hat{b}_2 = -5244.21733. \tag{B.14}$$

Substituting values b_1, b_2, N_1, N_2 in Equation (B.3), we get an algebraic equation

$$f(x) = 1.519x^{453} + 0.122x^{235} - 5.5 \times 10^{-9}x^{218} + 10^{-4}x^{17} - 1.7 \times 10^{-7} = 0 \tag{B.15}$$

Since the condition $N_2 > 2N_1$, Equation (B.15) has either one or three roots. At the inflection point the second derivative of $y(t)$ changes sign, i.e. convexity of the $y(t)$ function is replaced by a concave condition or vice versa. Suspicious in this regard are points of the $y(t)$ graph with abscissa $t_0^{(2)} = 36$ and $t_0^{(1)} = 24$.

We solve the equation with the initial condition, $x_0 = e^{-\tau_0}$, $\tau_0 = t_0/N$, where $t_0 = t_0^{(j)} (j = 1, 2)$. Calculations show that for any of these initial conditions, the equation has the same solution $x^* = 0.687$. Consequently, function $y(t)$ has one inflection point $t^* = N \ln(1/x^*) = 37.528$ or, going back to the original time scale, we obtain $t = 0.37528$.

References

1 Bendat, J.S. and Piersol, A.G. (2010). *Random Data: Analysis and Measurement Procedures*. Wiley.
2 Volterra, V. (1931). *Leçons Sur la Théorie Mathématique de la Lutte Pour la Vie*. Paris: Gauthier Villars.
3 Neimark, Y.I. and Landa, P.S. (1992). *Stochastic and Chaotic Oscillations*. Springer Science +Business Media Dordrecht.
4 Lichtenberg, A.J. and Lieberman, M.A. (1992). *Regular and Chaotic Dynamics*. New York: Springer-Verlag.
5 Moon, F.C. (2004). *Chaotic Vibrations: An Introduction for Applied Scientists and Engineers*. Hoboken, New Jersey: Wiley.
6 Postan, M.Y. (1993). Generalized logistic curve: Properties and estimate of parameters. *Economics and Mathematical Methods* **29** (2): 305–310.

Index

a

applications in production operations
  cement property enhancement   134
  high performance of self-compacting concrete (HPSCC)   133
  hydrogen sulfide removal   135
  nanoemulsions   134
  nanomaterial-based drilling fluid   134
  oil-based mud applications   135
  reduction of penetration rate (ROP)   134
  self-healing behavior   134
  well cementing   133

c

cement stone strength
  nano-fillers positive effect   137
  Portland cement
    concentration of TiO_2 and SiO_2 138
    granulometric composition   138
    X-ray diffraction analysis   139
  regression equation   141–142
  surface area and distribution of particles   139–140
  TiO_2 and SiO_2 nanoparticles   140–141
colloidal dispersion nanogels
  aging effect   200
  interfacial tension   200–202
  particle size distribution   203–204
  resistance factor/residual resistance factor   204–206
  rheology   198–200
  study and research methods   197
  TiO_2 nanoparticles   198
  viscosity   198–200
  zeta potential   202–203
colloids
  classifications of
    associated colloids   4
    lyophilic/intrinsic colloids   3
    lyophobic/extrinsic colloids   4
    macromolecular colloids   4
    multimolecular colloids   4
  defined   3
  evaluation   4–5
  filtration   3
  Tyndall effect   3
core flooding experiments
  dispersion medium properties   175
  dispersion phase properties   174
  IFD process, PPGs   192–196

d

Derjaguin–Landau–Verwey–Overbeek (DLVO) theory   9–10
determination of inflection points   263–265
dispersed-phase particle interaction   259–262

e

elastic–steric interaction   8
electrostatic interactions   7
enhanced oil recovery (EOR) method
  core flooding experiments
    dispersion medium properties   175
    dispersion phase properties   174
    nano-dispersion phase impact   173
  Extended Metal Atom Chains (EMACs)   227

Nanocolloids for Petroleum Engineering: Fundamentals and Practices, First Edition.
Baghir A. Suleimanov, Elchin F. Veliyev, and Vladimir Vishnyakov.
© 2022 John Wiley & Sons Ltd. Published 2022 by John Wiley & Sons Ltd.

f

fluidization agent
  chemical additives impact
    polymer compositions   152–153
    water–air mixtures with surfactant additives   151–152
  gasified fluids
    carbon dioxide gasified water   146–149
    natural gas/air gasified water   149–151
    sand plug washout   145

g

gas condensate mixture flow
  mechanism   111–124
  phase diagram   112
  porous medium wettability mechanism   120–121
  pressure build-up mechanism   121–123
  rheology mechanism   112–120
    annular flow scheme   113–115
    slippage effect   115–120
  volumetric flow rate   111
gasified Newtonian liquids
  nanogas emulsions hydrodynamics   41–60
    annular capillary flow scheme   41–44
    porous media   52–60
    slip effect   44–51
  porous media at reservoir conditions
    apparent permeability   53–54
    fundamental equations   52–53
    steady-state flow   54–60
  slip effect
    annular liquid/gas flow   45
    bubble volume   50
    capillary volume   50
    Navier hypothesis of the velocity   44
    near-wall layer thickness at different slip coefficients   47
    partial surface coverage   47
    relative liquid flow rate at different capillary radii   45, 48–50
    relative near-wall layer thickness   48
    slip coefficient at different degrees of pore channel surface coverage   47
    variable viscosity   46
  steady-state radial flow   73–81
    dimensionless flow rate   75, 76, 79, 81
    Dupuit flow rate   74
    permeability of matrix blocks   75
    pressure distribution   78, 80
    pressure–flow rate dependences   77–78, 80
    quasi-homogeneous incompressible fluid   73
    radial flow scheme   74
gasified non-Newtonian liquids
  nanogas emulsions hydrodynamics in heavy oil reservoirs   60–66
  porous media at reservoir conditions   66–73
  porous media at reservoir conditions
    capillary flow   66–67
    flow in heterogeneous porous medium   72–73
    flow in homogeneous porous medium   67–72
  slip effect
    maximum liquid flow rate   63
    Navier hypothesis   63, 65
    relative flow rate vs. boundary layer thickness   65–66
    relative liquid flow rate vs. pressure gradient   64
  steady-state radial flow   81–86
    Kozeny-Carman porous medium   81
    pressure distribution   85
    pressure–flow rate dependences   82–84

h

Hamaker constant   7, 9
hydrophobic and lipophilic polysilicon (HLP)   22
hydrophobic interaction   8

i

in-depth fluid diversion (IFD) process, PPGs
  bulk gel injection   188
  carboxyethyl cellulose (CMC)   188
  core flooding experiments   192–196

kinetic mechanism of gelation 191–192
nanogel strength evaluation 189–191

l
lipophobic and hydrophilic
 polysilicon (LHP) 22

m
magnetic dipole-dipole interaction 8
metal string complex (MSC)
 EMACs 227
 homonuclear MSCs 228
 Krogmann salt 227
 thermal conductivity enhancement of microfluids with$Ni_3(\mu3\text{-ppza})_4Cl_2$, 228–229
  fluid stability 230–231
  freezing point 236–237
  microparticles 229–230
  Ni_3 microfluid 230
  rheology of 235
  surface tension 235–236
  thermal conductivity 232–235
 thermophysical properties of nano-and microfluids with $Ni_5(\mu5\text{-ppza})_4Cl_2$ 237–238
  fluid stability 240–242
  freezing points 249
  micro-and nanofluid preparations 239–240
  microparticles 238–239
  rheology of 245
  surface tension 245–249
  thermal conductivity 242–245
  thixotropic colloidal structure 247

n
nanoaerosoles, gas condensate flow
 gas condensate mixture flow mechanism 111–124
  phase diagram 112
  porous medium wettability mechanism 120–121
  pressure build-up mechanism 121–123
  rheology mechanism 112–120

 in a porous medium 107–111
  compressibility coefficient 111
  experimental setup 107–109
  gas flow rate 110
  hydraulic diffusivities 110
  pressure build-up curves 110
nanocolloids
 defined 5
 depletion interaction 8
 elastic–steric interaction 8
 electrostatic interactions 7
 hydrophobic interaction 8
 magnetic dipole-dipole interaction 8
 osmotic repulsion 9
 particle aggregate 4
 particle structure 5
 polydispersity 5
 reservoir conditions, in-situ formation
  nanoaerosoles 38–40
  nanogas emulsions 35–38
 rheology of
  nanoparticle interaction 10–14
  nanoparticle migration 14–21
 solvation interaction 8
 stability of 9–10
 surface tension 23–24
 types of 4, 6
 Van der Waals interactions 7
 wettability
  defined 23
  hydrocarbon production 22
  interfacial energy 21
  of nanoparticle intrusion 21
  polysilicon 22
  porous media 21
  water-based fluid system 23
nanogas emulsion
 oil displacement
  nanogas emulsion injection 213–215
  Newtonian gasified fluid 207–208
  non-Newtonian gasified fluid 208–210
  slippage effect 210–212
 sand control
  anionic surfactant additive 155

nanogas emulsion (cont'd)
  chemical additives impact on
    fluidization 151–153
  fluidization by gasified fluids 145–151
  observed phenomena kinetic
    mechanism 153–155
  relative granular material
    expansion 152–153
  vibrowave stimulation
    approximate solution 161–162
    Duhamel integral equation 158
    elastic waves 158, 159
    exact solution 158–161
    mechanism of 157
    piezoconductivity equation 157
    slip effect 160
    transcendental equation 160
nanogas emulsions hydrodynamics
  gasified Newtonian liquids 41–60
    annular capillary flow scheme 41–44
    porous media 52–60
    slip effect 44–51
  gasified non-Newtonian liquids
    in heavy oil reservoirs 60–66
    porous media at reservoir
      conditions 66–73
nanoparticle interaction on colloid rheology
  dilatant flow 12–13
  dispersion flow curve 11
  flow cell transparency 11
  fractal dimensionality 13–14
  hydrolyzed PAN solution 13
  nonlinear effects 10
  power spectral density 12
  pseudoplastic flow 11, 13
  sheared dispersion 14
  thixotropic process 14
nanoparticle migration on colloid rheology
  clay slurry flow rate 19
  dimensionless apparent viscosity 16
  Einstein equation 17
  experimental setup 15
  exponential law 14
  Gryazevaya Sopka and Busachi
    deposits 15
  liquid flow rate 18, 20

non-Newtonian oil 15
qualitative conformation 19
rheological curve for the oil sample 16
natural gas composition 108
neutral wettable polysilicon (NWP) 22
Newtonian oil displacement
  heterogeneous porous medium 184
  homogeneous porous medium 184, 185
  oil flow rate 186
  oil recovery curves 185
  test process 183
  water-free and finite factors 185

o

oil displacement, nanogas emulsions
  nanogas emulsion injection 213–215
  Newtonian gasified fluid 207–208
  non-Newtonian gasified
    fluid 208–210
  slippage effect 210–212
osmotic repulsion 9

p

pendant drop method 245
porous media at reservoir conditions
  gasified Newtonian liquids
    apparent permeability 53–54
    fundamental equations 52–53
    steady-state flow 54–60
  gasified non-Newtonian liquids
    capillary flow 66–67
    flow in heterogeneous porous
      medium 72–73
    flow in homogeneous porous
      medium 67–72
Portland cement
  concentration of TiO_2 and SiO_2 138
  granulometric composition 138
  X-ray diffraction analysis 139
preformed particle gels (PPGs), IFD
  bulk gel injection 188
  carboxyethyl cellulose (CMC) 188
  core flooding experiments 192–196
  kinetic mechanism of gelation
    191–192
  nanogel strength evaluation 189–191

s

sand control, nanogas emulsion
  anionic surfactant additive 155
  chemical additives impact on fluidization
    polymer compositions 152–153
    water–air mixtures with surfactant additives 151–152
  fluidization by gasified fluids
    carbon dioxide gasified water 146–149
    natural gas/air gasified water 149–151
    sand plug washout 145
  observed phenomena kinetic mechanism 153–155
  relative granular material expansion 152–153
slip effect
  gasified Newtonian liquids
    annular liquid/gas flow 45
    bubble volume 50
    capillary volume 50
    Navier hypothesis of the velocity 44
    near-wall layer thickness at different slip coefficients 47
    partial surface coverage 47
    relative liquid flow rate at different capillary radii 45, 48–50
    relative near-wall layer thickness 48
    slip coefficient at different degrees of pore channel surface coverage 47
    variable viscosity 46
  non-gasified Newtonian liquids
    maximum liquid flow rate 63
    Navier hypothesis 63, 65
    relative flow rate vs. boundary layer thickness 65–66
    relative liquid flow rate vs. pressure gradient 64
slippage effect
  steady-state radial flow 73–86
  unsteady state flow 86–95
  viscosity anomaly, phase transition point 95–106
steady-state flow
  dimensionless liquid flow rate 56–57, 59
  flow flux 55
  gas condensate flow 120–121
  gas saturation 59
  liquid flow rate 55
  pressure distribution 54, 58, 60
  rheology mechanism 112–120
  wettability 57
steady-state radial flow
  gasified Newtonian fluid flow 73–81
    dimensionless flow rate 75, 76, 79, 81
    Dupuit flow rate 74
    permeability of matrix blocks 75
    pressure distribution 78, 80
    pressure–flow rate dependences 77–78, 80
    quasi-homogeneous incompressible fluid 73
    radial flow scheme 74
    slippage effect 73
  gasified non-Newtonian fluid flow 81–86
    Kozeny–Carman porous medium 81
    pressure distribution 85
    pressure–flow rate dependences 82–84
subcritical gas nuclei, stability
  Boltzmann's law 36
  morphology of 38
  nucleation process 35
  threshold radius 37
  Tolman's equation 36
subcritical liquid nuclei, stability 39–40
surface tension 23–24
surfactant-based nanofluid
  adsorption process 178–179
  Newtonian oil displacement 183–186
  on oil wettability 179
  optical spectroscopy results 179–182
  rheological properties of nanosuspension 182–183
  on surface tension 177–178

t

thermal conductivity enhancement of microfluids with $Ni_3(\mu3\text{-ppza})_4Cl_2$, MSCs 228–229
  fluid stability 230–231

thermal conductivity enhancement of microfluids with $Ni_3(\mu3\text{-ppza})_4Cl_2$, MSCs (cont'd)
    freezing point 236–237
    microparticles 229–230
    Ni_3 microfluid 230
    rheology of 235
    surface tension 235–236
    thermal conductivity 232–235
thermophysical properties of nano-and microfluids with $Ni_5(\mu5\text{-ppza})_4Cl_2$, MSCs 237–238
    fluid stability 240–242
    freezing points 249
    micro-and nanofluid preparations 239–240
    microparticles 238–239
    rheology of 245
    surface tension 245–249
    thermal conductivity 242–245
    thixotropic colloidal structure 247
transient hot wire method (THWM) 242
Tyndall effect 3

u

unsteady state flow
    bottomhole pressure 89
    boundary conditions 86, 88
    diffusion gas dissolution 94
    flow time dependence 88
    gas condensate flow 121–123
    initial conditions 86
    pressure build-up curves 90–95
    pressure build-up time 90, 92, 95
    slippage effect 86, 87

v

Van der Waals interactions 7
vibrowave stimulation, nanogas emulsion
    approximate solution 161–162
    depth of penetration 163

Duhamel integral equation 158
elastic waves 158, 159
exact solution 158–161
mechanism of 157
piezoconductivity equation 157
slip effect 160
transcendental equation 160
viscosity anomaly, phase transition point
    experimental procedures 96–97
        Karpatol-UM2K-Nurol 96–97
        live oil sampling 96
        Model 3000 GL Chandler Engineering PVT Phase Behavior System 96–97
        oil properties 96
    live oil viscosity measurement 97
    mechanism of 102–105
    phase behavior of live oil 97–99
    live oil samples 99
    live oil viscosity dependence on pressure 97, 99, 101
    mechanism of surfactant influence on 105–106
    oil surface tension on surfactant concentration 105
    pressure–volume plot 100
    saturation pressure on surfactant concentration 100
    surfactant impact on 100–102

w

water alternating gas (WAG) method 24
wettability
    defined 23
    hydrocarbon production 22
    interfacial energy 21
    of nanoparticle intrusion 21
    polysilicon 22
    porous media 21
    water-based fluid system 23